FEEDING THE FIRE

FEEDING THE FIRE

The Lost History and
Uncertain Future
of Mankind's
Energy Addiction

MARK E. EBERHART

HARMONY BOOKS
NEW YORK

Published in the United States by Harmony Books, an imprint of the
Crown Publishing Group, a division of Random House, Inc., New York.
www.crownpublishing.com

Harmony Books is a registered trademark and the Harmony Books
colophon is a trademark of Random House, Inc.

Library of Congress Cataloging-in-Publication Data
Eberhart, Mark E.
    Feeding the fire : the lost history and uncertain future of
mankind's energy addiction / Mark E. Eberhart. — 1st ed.
    Includes bibliographical references.
    1. Renewable energy sources—United States. I. Title.
  TJ807.9.U6E34   2007
  333.79—dc22   2006032707

ISBN 978-0-307-23744-6

Printed in the United States of America

DESIGN BY BARBARA STURMAN

10 9 8 7 6 5 4 3 2 1

First Edition

To all of those who wonder, "What if?"

Acknowledgments

I would like to express my sincere appreciation for all the encouraging letters and e-mail regarding *Why Things Break*. If not for these, it is unlikely I would have attempted a second book. I am grateful to Jim Hornfischer for his willingness to consider each of the many topics suggested as subject matter for this book. The crew at Harmony was great, and I am particularly thankful to Julia Pastore for her encouragement and patience. I would also like to acknowledge Carolyn Thomas's efforts. She read, reread, edited, and reedited. Thanks to Paul Jagodzinski for arranging a teaching schedule that made it possible to write. And finally, thanks to the Davis family—Kyle, Ann, and Parker—for providing me with a roof when I had none.

Contents

INTRODUCTION

Energy and Imagination   xi

NATURE'S ENERGY LAWS

1: The Thinking Man's Diet   3
2: Diet Basics I: Nothing But Energy   12
3: Diet Basics II: All Energy Is Not Equal   28
4: Talking Thermodynamics, Thinking in Pictures: Entropy   39
5: The Thermodynamics of Forgetting: The Energy
in Information   52

ENERGY METABOLISM

6: The Energy Ride: The Big Picture of Energy Metabolism   67

7: The Age of Atoms   80

8: The Age of Molecules   92

9: From Life to Fossil Fuels   106

10: Keep Cool, Man, Real Cool: Human Evolution and
the Origins of Imagination   121

OUR ENERGY PAST

11: The Great Energy Rule: First Came Agriculture   143

12: A Tale of Two Cities: Wood to Burn   151

13: Saving the Forests: Coal to Burn   163

14: And There Was Light: Oil to Burn   172

15: Power Unleashed: Electricity   194

OUR ENERGY PRESENT

16: First, the Bad News   215

17: Now for the Good News   226

OUR ENERGY FUTURE

18: The Thinking Man's Energy Diet   245

Notes   263

References   269

Index   273

Introduction

ENERGY AND IMAGINATION

Finding and exploiting energy is the human story.

One or two hundred thousand years ago, anatomically modern humans were born to this world. Their minds set them apart from all other creatures, for the spark of creativity resides there. That spark fell on combustible tinder, and the fire of imagination sprang to life. From that point onward, feeding the fire was as essential to sustaining our lives as was slaking our thirst or satisfying our hunger.

At first, we tended the small fire by inventing efficient ways to hunt and gather. Then, we turned to agriculture and used the work of animals as fuel for the fires of our imaginations. The fire grew and spread, until the labor of countless animals could no longer provide the energy demanded. So our imaginations enlisted the power of wind and water to grind grain and irrigate fields.

Still the fire grew.

We burned the wood of many forests for heat to smelt and forge metals, bake bricks, and make mortar and cement. With these, we built the roads, ships, and cities of our imaginations.

Still the fire grew.

We tunneled underground, harvesting energy-rich coal to drive steam engines, powering the supply line of factories, locomotives, and ships that produced and distributed the products of our imaginations.

Still the fire grew.

We diverted and dammed the world's largest rivers, using the expropriated energy to make chemicals, light cities, and run air conditioners. We drilled into the earth, drawing from primordial pools of oil to fuel tractors, trucks, and cars. We tapped the energy of the atom for power to digitize, archive, and retrieve the knowledge and insight of incalculable thousands.

And still the fire grows.

Our world pumps and then burns more than three and a half billion gallons of oil every day. A little less than one-fourth of that is burned right here in the United States, and though in the years prior to and immediately following World War II the United States was the earth's leading oil exporter, producing more than enough to satisfy domestic needs, today's situation is much different. We import more than half the oil necessary to feed our imaginations. And Americans tremble with fear when the oil-rich and politically unstable countries of the Middle East use the words *reduced oil production, quotas,* or *embargo.*

The first of two 1970-era oil shocks began in 1973, when the Saudis, who sit atop the world's largest proven oil reserve, were unhappy with U.S. policy toward Israel. To make their point, they delivered, right on target, the *oil weapon:* they embargoed all oil shipments to the United States. Domestic energy prices skyrocketed, insecurity raced through financial markets, and like viscera cut from America's soft underbelly, cars lined up at service stations awaiting gas deliveries.

For a brief time Americans responded, insulating their homes and turning to more fuel-efficient vehicles. U.S. oil consumption fell. Then, in 1980, Ronald Reagan was elected president. When he declared that conservation meant "being too hot in the summer and too cold in the winter," America was back on track to being the most wasteful country of all time. And though Reagan called on the scientists of the free world to design and build an antimissile system that would make the nuclear weapons of our enemies "impotent" and "obsolete," he proposed no plan or strategy to counter the oil weapon.

Those oil shocks were artificially induced: Oil production capacity exceeded demand, and thus the oil weapon required withholding supplies. The world of the twenty-first century has changed. China and India, with combined populations in excess of two billion, are now developing countries whose citizens are every bit as anxious to feed their imaginations as are Westerners. There is no longer significant excess production capacity, and the situation is expected to get worse as worldwide energy demand grows in the face of steady, or possibly shrinking, supplies. Some forecasters believe that this structural oil shock will cause upward-spiraling energy prices. Where will gas prices stop? Five, six, ten dollars a gallon?

How will we in the United States respond as the plentiful and abundant energy that has fed and nourished our imaginations becomes scarce? We treat energy as background noise, unaware of how dependent we have become. Whereas two hundred years ago a summer evening's activities might have brought the family together to share poetry by the light of a whale-oil lamp, today's kids grab burgers at drive-thru windows before zipping across town to the multiplex. Mom and Dad take the SUV to the gym, work out, take a steam, and then head home to watch a movie on pay-per-view. The pastimes of two hundred, fifty, even twenty years ago are now boring, because our imaginations have grown accustomed to a higher-energy diet, one that may not be sustainable.

Indeed, it is an unusual week when the national news does not include a story about the growing environmental consequences of unrestrained energy use. Predictions surrounding the dire consequences of environmental damage and global warming are rampant. Some of the more ominous warnings presage the melting of both the Antarctic and Greenland ice caps, a sea level increase of fifteen or more meters, desertification, intensifying tropical storms, declines in arable land, sinking world food production, expansion of tropical parasites to midlatitudes, and mass extinction.

On the heels of these predictions and warnings, the course ahead should be clear. Yet, with energy as the key resource allowing us to exercise our imaginations, it is not surprising that we are ambivalent as to how to proceed. We have developed a split personality: On one hand, we are alert to the consequences of unrestrained energy use; on the other, we live for the moment, making decisions based on emotion, regardless of cost or consequences.

And so we find ourselves in this amazing and unprecedented circumstance where our expanded imaginations lead us to envision a future where energy is no longer plentiful; a future where the atmosphere has warmed, creating widespread drought; a future where the ice caps have melted and our coastal cities are flooded. The image is a disturbing one, and we are motivated to take action—to stop burning fossil fuels and outlaw gas-guzzling SUVs. At the same time, we are unwilling to relinquish the energy that has nourished our minds, expanded our horizons, and improved our lives.

While pondering our energy future, it is worth remembering that this is not the first time humankind has confronted an energy crisis. Energy shortage is apparently part and parcel of cultural development. Some societies do not survive their brush with scarcity; others do. The Maya, Anasazi, and Easter Islanders used their energy resources to sustain complex cultures distinguished by magnificent art and architecture. When their energy resources were exhausted, these civilizations vanished, leaving behind only buildings and monuments to mark their passing. Perhaps they recognized their precarious circumstances and made futile plans to save their societies, but without an alternative energy source, their futures were doomed.

The cultures of medieval Europe were more fortunate. On the eve of the Industrial Revolution, the prime energy source of the time was under threat. Great parts of the European continent had been deforested for arable land, timber, and energy. Governments recognized the impending crisis but were up against their own insatiable appetites for navies and iron weapons to guarantee national security. The citizenry required steel tools, timber, glass, lime, and bricks for buildings, and beer and distilled spirits to satisfy creature comforts. For these, and other things, the clearing of timberland for energy could not be halted, or even slowed. Eventually, coal was forced on Europe as the fuel of last resort. Why is it that civilizations that faced collapse often could not take the necessary steps to conserve their energy resources? It is because they could not imagine an alternative energy future. We can.

As the forests of Europe fell to the woodsman's ax, a great debate raged among Newton, Descartes, and Leibniz—the most acclaimed scientists of the day—as to exactly what energy was. They called it "the living force." But when it came to quantifying energy, or measuring it, they were clueless. No one of the period recognized that work and heat were simply forms of energy, or that work could be converted entirely to heat—though heat could not be converted wholly into work. Thanks to scientists, philosophers, and engineers of the seventeenth, eighteenth, and nineteenth centuries, all of this is now common knowledge. We know the maximum work we can extract from heat and how to build machines that operate near this optimum efficiency. We know how to store and liberate the energy of chemical reactions, and how to harvest the energy held captive inside the nuclei of atoms. We know how to convert sunlight, wind, and waves into electricity, and how to transmit that energy over vast distances. We know how to turn coal into a liquid to fuel our cars. We know of vast supplies of "pre-oil" locked in rock and how to process it into the real thing. We are also aware of the environmental consequences of unrestrained energy use. We have developed technologies to scrub combustion air free of harmful contaminants before they enter the environment. We know that the accumulation of atmospheric carbon dioxide can cause global warming, and we have developed technologies to mitigate this possibility.

Whereas the people of medieval Europe had to settle for the energy future given them by happenstance and necessity, we possess the knowledge and technology to plan our future. What we appear to lack is the will to make such a plan.

While the leaders of the United States feel comfortable adopting a "policy of preemption" that accepts a military response "even if uncertainty remains as to the time and place of an enemy's attack," their policy toward the threat posed by energy dependency and global climate change is more of a wait-and-see approach. A National Academy of Sciences study concluded that the "balance of evidence" shows global warming effects are evident. Still, President George W. Bush has repeatedly insisted that additional scientific research is needed before he will impose restrictions on greenhouse gas emissions. As the Bush administration awaited "scientific certainty" for the cause and effects of global climate change, it found the evidence that Saddam Hussein possessed weapons of mass destruction sufficiently compelling to launch the preemptive invasion and occupation of Iraq. Some analysts suspect the real cost of that war will exceed two trillion dollars (not to mention the loss of lives on all sides, which in 2006 may surpass 500,000). Though no WMDs were found, it seems that the mere suspicion—unsupported by any hard data—of peril that could "come in the form of a mushroom cloud" justifies these expenses.

Set aside the issue of global warming. What of energy dependency? It shouldn't come as a surprise to learn that Osama bin Laden has stated that incapacitating the U.S. economy is a primary al-Qaeda objective—placing oil markets and global energy infrastructure high on its target list. Thus, Washington is in a position of defending the oil interests of countries that are politically and philosophically at odds with our historical position on human rights. Further, our dependency makes it difficult to oppose the political and military ambitions of these countries. Iran, for example, thumbs its nose as the rest of the world struggles to limit Tehran's nuclear ambitions. To limit dependency, the Bush administration proposes further development of domestic sources of crude, including those in the Arctic National Wildlife Refuge (ANWR). The most optimistic estimates of this reserve's production is just under a million barrels per day—the same amount of oil that would be saved by increasing overall automobile and light-truck fuel efficiency by between two and three miles per gallon.

The issue, of course, is neither improved fuel efficiency nor drilling in ANWR. It is not war with Iraq or recognizing global warming as an environmental, economic, and security threat. It is the fact that we do one without even considering the other. We choose war, and we choose *not* to explore technological approaches to limit greenhouse gas emissions. We choose to promulgate drilling in ANWR, but *not* to support research and development to increase fuel efficiency. Our leaders refuse to embrace science and technology as tools to achieve economic and political ends. They see the world as it was fifty or a hundred years ago, when disputes were settled with conflict and shortages were satisfied by further exploitation. Our politicians make it apparent that, for them, understanding science and using technology to solve problems is too hard.

Contrast today's leadership with that of John F. Kennedy as he spoke of his vision of the future in his September 12, 1962, call to the nation to put a man on the moon by the end of the decade:

We choose to go to the moon in this decade and do the other things, not because they are easy, but because they are hard, because that goal will serve to organize and measure the best of our energies and skills, because that challenge is one that we are willing to accept, one we are unwilling to postpone, and one which we intend to win, and the others, too.

What has changed? Why did we see science and technology as the means to "organize and measure the best of our energies" and now we discount and ignore the very capabilities that these pursuits have provided? The solution to what is arguably the most pressing problem that has faced the people of the United States, if not the world—a renewable and sustainable energy future—lies in exploiting the knowledge we have developed over millennia. Thirty years from now the world's superpowers will not be those with the most powerful militaries; they will be the countries that imagine an efficient energy future and apply the world's knowledge to make this future real.

If the United States is to remain a superpower, we must recapture a sense of confidence in what we can accomplish with our most abundant and valuable resource: *our imaginations.*

Nature's Energy Laws

THE THINKING MAN'S DIET

At twelve, I was convinced I had stumbled upon the ultimate diet, and it was going to make me rich. I was a skinny kid who spent most of my adolescence combining Hostess Twinkies, instant pudding, heavy cream, lots of eggs, and cans of sweetened condensed milk in various high-caloric proportions in a tireless effort to bulk myself up—while the rest of my family struggled with their weight.

My stepfather swore by *The Drinking Man's Diet,* the invention of Robert Cameron (aka Jeffrey Roberts), a San Francisco bon vivant and entrepreneur who marketed a "fun" diet in a fifty-page pamphlet that sold for one dollar in 1964. Predating Atkins by nine years, Cameron advocated healthy weight loss by reducing one's consumption of carbohydrates, leaving dieters free to dine on porterhouse steaks, lobsters languishing in garlic butter, and salads smothered in Roquefort dressing. Cameron's real genius was in recognizing that distilled spirits like vodka, gin, and whiskey contain only trace amounts of carbohydrates. Hence, there was no reason to abstain from a brandy after that Chateaubriand. In short, Cameron preached that it was possible to lose weight while eating, drinking, and making merry. So appealing was this message that in only two years' time *The Drinking Man's Diet* had been released in thirteen languages and sold 2.4 million copies. Cameron was set for life. At this writing, the still svelte ninety-something's diet book is still in print.

If Cameron could get rich by inventing a diet, why couldn't I? In addition to a diet simply working, it seemed a catchy name was essential to success. I would call mine *The Thinking Man's Diet.*

Somewhere—I can't remember where—I had learned that our brains consume more energy than our skeletal muscles. It was obvious: Think more and lose weight. I pictured a daily regimen where calisthenics were replaced with solving algebra problems, jogging with games of chess, and pumping iron with deciphering the daily crossword puzzle. And if you ate that dish of triple-chocolate-fudge ice cream after dinner, a few jumbles or geometric proofs before bed would serve as penance. This wonderful fact also seemed to explain why I was so skinny—I just thought too much, which is exactly what my grandmother always said. It was clear, at least to me, that *The Thinking Man's Diet* could make millions.

But there was a glitch in my master plan. Specifically, the relative energy consumption of brains versus muscles is with respect to the basal metabolic rate, or BMR. For persons totally at rest, their brains consume energy faster than their muscles. Working muscles, however, are a different story.

The skeletal muscles of a typical 75 kg (165 lb) adult male make up about 40 percent of his body mass. This resting muscle consumes slightly less energy than a 15-watt refrigerator bulb. But put only some of these muscles to work pedaling a bicycle and a moderately conditioned rider will burn energy at the same rate as five 100-watt lightbulbs. A world-class rider can use energy at the same rate as a handheld hair dryer, 2,000 watts.

So, if algebra were to substitute for calisthenics, a thinking brain would need to use much more energy than one at rest. Here was the flaw in my get-rich-quick diet plan: Once again, consider a 75 kg adult male whose brain will account for approximately 2 percent of his body mass, roughly 1.5 kg. While sleeping, this brain will be using slightly more than 15 watts. Wake our subject and confront him with a calculus exam, and his brain will still require 15 watts of power. Have him solve the *New York Times* Sunday crossword puzzle, take an IQ test, or learn and explain the General Theory of Relativity, and how much power will his brain use? You got it.

Our brain is the only organ that consumes energy at a nearly constant rate, while all other organs respond to the demands placed upon them. Our hearts, which account for only one-half of one percent of our body mass, use a whopping 10 watts of power while resting. Ask the heart to supply blood to the muscles of our cyclist, and its power requirements will increase. The same is true of our kidneys, liver, organs of the gastrointestinal tract, and lungs—all use energy in proportion to the work they are being required to do. Except for our brains.

Alas, there was to be no *Thinking Man's Diet.* There was to be no pamphlet—not in any language, let alone in thirteen of them.

The adage "You can do anything you put your mind to" had to be nonsense. With a constant power supply, our brains must function like a car with a broken accelerator—the driver can neither speed up nor slow down. Is this true? Are we limited in what we can do with our brains? We can check by measuring the rate at which *brainwork* accumulates. If our brains are like that car with the broken accelerator, brainwork will accumulate at a constant rate.

However, before we can start our bookkeeping, we need a more tangible definition. What exactly is the product of the working brain?

For most organs, this is an uncomplicated question. The work done by the heart is measured by the quantity of blood it pumps. The work done by the gut is related to the amount of food digested. And the work done by the lungs is associated with the volume of air inhaled. When it comes to the brain, though, the issue is not as obvious.

An essential component of brainwork is connected with actuating the nerves of the voluntary and involuntary muscles. When I reach across the table to grab the last chocolate chip cookie, it is because my brain is sending signals to my arm, shoulder, hand, and fingers. However, in human beings, only a small portion of the brain is involved with this task; the greater part is busy with other activities that are more easily illustrated than explained. So, picture yourself driving a rollover-prone SUV at 75 mph on a straight stretch of interstate. Suddenly, a piece of debris causes the front right tire to blow.

Two things have just occurred: First, you did exactly as I asked. You formed a mental image of an SUV striking a piece of debris. Second, not only did you imagine the SUV as instructed, but you also took the next step unbidden and pictured the vehicle tumbling end-over-end down the interstate. Our imaginations display an almost physical inertia—once the process begins, it must continue. This is what imagination is all about. We have the ability to hold on to real or fictitious ideas and events and explore the multifold possible outcomes that may result. In short, we have the ability to ask and seek answers to the question "What if?" Your response to the SUV situation is the natural result of attempting to envision the consequences of the question I posed, "What if the SUV you are driving has a blowout?"

For the most part, we are unaware of all the *What if?* questions our minds grapple with on a daily basis. All of us have had at least one of those *Eureka!* moments, when an idea or solution to a problem seems to pop into our heads. In fact, when this happens our brains have been struggling with the problem for hours, days, perhaps even years, experimenting with one scenario after another. It is only after (finally) settling upon a solution or course of action that the question is shifted from the mental background to consciousness—and, *Eureka!* Our brains are always questioning, always searching for answers, always imagining. Our brains are like that car with the broken accelerator, always running at full speed.

Yet searching, questioning, and thinking are not brainwork, any more than revving the engine of a parked car is driving. Work is the useful product of an action. Whereas our heart, lungs, and gut do work by maintaining our body, our brain does work when providing the stuff that can nourish *another* person's imagination. You don't have to look far to find examples of brainwork. This book was written to nourish your mind. In turn, I was only able to write it because I had access to a computer, text-editing software, the Internet, and reference books. All of these things served to feed my imagination. And the designers of each of these things drew sustenance from the inventions and ideas of others—from other inventors' and scientists' and writers' brainwork.

Brainwork is self-propagating. Each invention grows our collective imagination. With a more robust imagination, we create even more things, which further nourish our imaginations, and so on. The adage about doing anything you put your mind to appears to be true after all. Which means we are left with a paradox: How can we do more with an apparently constant energy supply?

The solution is obvious. All of the "stuff" that nourishes our imaginations comes with an energy source. The energy to power the computer and text-editing software I mentioned did not come from the 15 watts allocated to my brain. I also did not have to draw from my own brain when looking for information—I did an Internet search that was powered through a worldwide electrical distribution system. And when I could not find what I wanted on the World Wide Web, I did it the old-fashioned way and read books, which had been produced with the aid of electric printing presses and were then distributed to libraries and bookstores with trucks powered by gasoline and diesel engines. In a very real sense, I could do more brainwork because my imagination was able to tap in to a virtually unlimited energy resource—a pipeline of brainwork, so to speak.

Brainwork is the defining product of a civilization. Hence, it is not surprising to see cultures flourish when energy is available and collapse when it dwindles. The United States—blessed with abundant timber, coal, oil, and hydroelectric power—has risen to become one of history's great civilizations. Our brainwork is unprecedented, including great works of art and literature, architecture and science. We have deciphered the human genome, cured diseases, and journeyed to planets. Yet, the very abundance that made this possible may now be failing.

Forty years ago, we produced the energy we used. Now we are dependent on other countries to provide the fuel demanded by the fires of our imaginations. Though we represent less than 5 percent of the world's population, we consume 25 percent of its energy production, importing over half of the more than 20 million barrels of oil we consume each day.

Other nations have instigated wars to win freedom from the yoke of energy dependency. December 7, 1941, is remembered most often as the day of the "unprovoked" attack on Pearl Harbor and the beginning of U.S. involvement in World War II. We seldom recall that before this date, America—the world's largest oil producer of the time—had embargoed oil shipments to the Japanese. The bombing of the U.S. fleet protected Japan's flank as its forces moved to secure the oil fields of Indonesia. And though our leaders deny that the war in Iraq has anything to do with oil, the United States is occupying the country with the world's second-largest proven reserves (112 billion barrels) and with undiscovered oil assets estimated at 200 billion barrels.

At the same time our reliance on foreign oil is growing, we have come to recognize that using the energy sources that have fed our imagination for centuries—coal, oil, and gas—may endanger the planet. The evidence for global warming has grown to the point that it is difficult for reasonable people to be dismissive of those alarmed by the prospect of climate change. These people envision a world that is far less hospitable than the one in which we presently live.

Confronted with energy dependency and global climate change, our course forward should be obvious; however, with energy as the key resource allowing us to exercise our imaginations, it is not surprising we are ambivalent as to how to proceed. We have developed a Jekyll and Hyde mentality. On one hand, like Dr. Jekyll, we strive to be responsible; on the other, like Mr. Hyde, we seek immediate gratification regardless of cost or consequences.

Dr. Jekyll has led us to consider energy in a more thoughtful way. Where not so long ago we concerned ourselves with the form of the energy we used, now it's how the energy is made that is more important. Hence, we speak of solar, geothermal, wind, hydroelectric, chemical, or nuclear energy. Of these, solar and wind are forms of renewable energy, which are considered to be good energy or "green" energy, indicating that no harm was done to the environment during its production. Hydroelectricity is a renewable resource that may be considered bad energy if it is generated in dams that interfere with salmon migrations, destroys stretches of whitewater, or obliterates mountain goat habitats. Since the accidents at Three Mile Island in 1979 and Chernobyl in 1986, nuclear energy has been firmly ensconced in the "bad" category, though geothermal energy, which is also a form of nuclear energy, is considered good. Chemical energy, commonly generated by burning something, can be good or bad depending on what is being burned. Coal-derived energy is bad, for example, whereas energy derived from ethanol is good. Ethanol is a form of biomass energy. It does not alter the amount of carbon dioxide in the atmosphere and therefore does not cause global warming. Though wood is also a form of biomass, the energy derived from wood burning is bad because it pumps smoke and other noxious substances into the air.

Mr. Hyde venerates energy consumption and makes his most important energy decisions based on *power*. Power is a measure of how fast energy is used. A modest car, when speeding away from a stoplight, is consuming energy as fast as a herd of 200 horses thundering across the open range. A stock H2 Hummer generates a paltry 316 hp compared with its cousin, the Viper, which churns out 500 hp. Its energy doesn't come from grass, however. One pours gasoline right down the throat of the V-10 engine powering this extraordinary auto. Both the Viper and the more common cars on the road, such as light trucks and SUVs, consume energy far faster than what is required for Mr. Hyde to simply get to and from work. But in his imagination, he sees himself whipping around the oval at Daytona or barreling up a 30 percent grade to some off-road Shangri-la.

In many ways, the ambivalence we suffer while trying to plan our energy future is no different from the hesitation people experience when starting a diet. Potential dieters want to be healthier, stronger, and more attractive but worry about giving up the foods they enjoy. What is needed is an energy diet like Cameron's drinking diet— one that will let us enjoy all the world has to offer without undue deprivation.

There has never been a better time to formulate such a diet, for we have spent centuries studying energy. We have at our disposal the brainwork of thousands; some from scientists who discovered the natural laws that control energy and its use. Others built on this brainwork, enabling us to build efficient machines and, more important, routes toward their improvement. We know how to extract energy from the wind, water, sea, and air. And we have come to appreciate the complexities of the environment and our interaction with it. We can imagine and make real a renewable and sustainable energy future if we can only gather the determination.

Like any good diet, motivation to persist must come from envisioning an energy future in which the United States will be politically and economically more secure, where our nation will be more competitive, improve its world image, and boost the standard of living for its citizens. And, most important, where we will be able to feed the fire of our imaginations without threatening our existence.

The diet I propose will be built on sound scientific footing. Though it will be scientific, it will not be abstruse. In fact, the background people will need to make sound energy decisions is astonishingly straightforward. It requires a consciousness of the two great laws of nature—the First and Second Laws of Thermodynamics—to which we are all subject. These laws cannot be broken—hence many tend to be unaware of their existence. This is a problem, for when politicians or bureaucrats are oblivious to nature's laws, nothing prevents them from formulating policies that have no possibility of achieving the desired result.

For our energy diet, becoming conscious of nature's laws is a little like recognizing the difference between carbohydrates and protein. It's a good starting point, but much more is required. We need to explore energy in detail—how it was formed and how it evolves. We will also strive to understand how we became energy-dependent creatures, and how this hunger has been satisfied for thousands of years. We will take a close look at the consequences of our energy appetite. Finally, we will put it all together to imagine a secure and renewable energy future and formulate a "diet" to achieve this end. Let me suggest: *The Thinking Man's Energy Diet.*

2

DIET BASICS I

Nothing But Energy

Having been fascinated with weight loss since an early age, I have come to believe that metabolism is key to a successful diet plan. This also holds true for an energy diet. There are two parts to our study of metabolism: energy and nutrients. We have all heard that junk food provides energy but is short on nutrition, and that for some unknown reason less appetizing green things, particularly Brussels sprouts and broccoli, are highly nutritious.

My mother had her own way of distinguishing between energy and nutrition. She called foods high in energy but lacking in nutritional value "empty calories" and those that were nutritional "good calories." When I was young and could consume eight to ten bottles of

Tattered Cover
Book Store

Books Are
Humanity
in Print.
—Barbara Tuchman

Coke a day without gaining an ounce, I was running on empty calories. Now that weight comes a little easier, I drink Diet Coke without any calories, empty or otherwise.

To formulate The Thinking Man's Energy Diet we must be cognizant of metabolism. It will not do to reduce our energy intake if the result is a malnourished imagination. For, just as there are empty calories that don't help our bodies stay healthy, there are "empty" forms of energy that can't be used for brainwork. Fortunately, two laws of nature will aid us in our examination of energy metabolism.

These laws form the joint columns upon which the entire structure of science—physics, chemistry, biology, physiology, and engineering—rests. The first of these great laws warns us against fad diets and quick fixes. It does not allow energy to be created or destroyed. In other words, energy is conserved. The second distinguishes between "empty" and "good" energy. We must spend a little time becoming better acquainted with these laws before we can tackle energy metabolism; and in deference to its name, I will begin with the First Law of Thermodynamics.

Before energy was called energy, it was called *vis viva,* the living force, and was thought of as what made the universe work. Its most obvious characteristic was making things move: stars, planets, people, horses, clouds, you name it. If it moved, it was endowed with *vis viva.*

The energy of motion, or what we now call *kinetic energy,* is but one of energy's many forms. It holds a special place in our discussion, for it was through motion that humankind was first alerted to the permanence of energy, and this realization grew from our fascination with moving things. The sling, throwing stick, blowgun, and bow and arrow were among the first man-made objects and did nothing more than impart motion to sticks and rocks.

But it didn't stop there. If something, anything, could be made to move, we found a way to do it. We invented catapults to hurl enormous stones and sails and oars to move boats. We harnessed animals to carts and moved our belongings. The Chinese invented rockets and Europeans altered this technology to make cannons and guns. Steam engines were invented to move water but were quickly modified to move riverboats and locomotives, which, in turn, moved us. That brings us to the "people movers": horses, cars, trucks, bicycles, unicycles, motorcycles, scooters, Segways, boats, submarines, planes, all-terrain vehicles, pogo sticks, Rollerblades, rickshaws, ice skates, skis, skateboards, snowmobiles, and parachutes—to name a few. And when not moving ourselves, we are frequently moving other objects. In fact, we have invented a plethora of things that serve no purpose other than to move: golf balls, tennis balls, footballs, baseballs, hockey pucks, kites, yo-yos, hula hoops, bowling balls, darts, Frisbees, boomerangs, and batons.

We are so enthralled by motion that those who move the fastest, farthest, or longest are enshrined. Why else would anyone try to set—let alone break—the land speed record of 762 mph? Or the 152-mph land speed record while motor pacing (where a bicycle drafts in the wake of a motorized vehicle)? Given our attraction to motion, it is clear why we are so easily seduced by the quick-fix diet of perpetual motion and a never-ending source of energy.

Claims for the discovery and invention of perpetual motion machines have a rich history. The earliest known account dates to the "magic wheel" of eighth-century Bavaria, which was supposed to exploit the magnetism of lodestones, causing a wheel to rotate forever. It didn't work. Five hundred years later, the French architect Villard de Honnecourt proposed mounting seven heavy movable hammers to the rim of a large wheel revolving around a stationary axle. When a hammer reached the apex of its rotation, it was hinged to fall forward in the direction of rotation. In this way, Honnecourt reasoned, there would always be more weight on the descending rather than the ascending side of the wheel, and hence its motion should be permanent. The bothersome fact that the device did not work hardly moderated the architect's enthusiasm. He advocated its use for such things as sawing wood, raising weights, and building "an angel whose finger turns always toward the sun." No less an intellect than Leonardo da Vinci made several drawings of machines he hoped would move eternally. A few years later, in 1518, Mark Anthony Zimara designed a windmill propelled by bellows powered by the windmill itself.

The list does not slacken in modern times. Guido Franch was convicted of defrauding his investors in 1973 with green powder that, when mixed with water, was supposed to transform into a high-octane fuel. In 1977, Arnold Burke raised almost $800,000 to support his development of a self-acting pump. During its demonstration investigators found a hidden electrical power source, paving the way for a fraud conviction and $250,000 in fines and penalties. So common are these continuing claims for perpetual motion that, as far as the U.S. Patent and Trademark Office is concerned, they fall into a special class that requires submitting a working model along with the standard patent application. What motivates the continuing hunt for the Holy Grail of machines, and the gullibility of those who invest in the search, is the absence of any proof that they do not exist. The impossibility of perpetual motion is rooted not in a mathematical truth, like the formula for the circumference of a circle, but in experience. Over millennia, every single attempt to build ever-moving machines has ended in frustration and failure. Most scientists have come to believe that our universe obeys natural laws, some of which preclude perpetual motion. One of the implications of this perceived truth is that energy is conserved. One cannot build a perpetual motion machine that would do work, like the type proposed by Villard de Honnecourt, because this would create energy.

The conservation of energy is a principle known to virtually every schoolchild. However, knowing and applying are two different things. The lure of perpetual motion is powerful—so much so that we can be misled to pursue policies founded on the faulty notion that energy can be created from absolutely nothing. This was the case when President George W. Bush unveiled his plans for a hydrogen-powered car during his State of the Union address of 2004. He proposed $1.2 billion in spending to develop a revolutionary automobile deriving power from hydrogen fuel cells. The president vowed that within twenty years, hydrogen fuel cell cars "will make our air significantly cleaner and our country much less dependent on foreign sources of oil."

With the benefit of becoming less dependent on foreign oil, Bush was proposing a program to create energy—in effect, a fad diet that wouldn't save a drop of oil. Why wasn't the *New York Times* running stories with headlines like "President Proposes Development of Perpetual Motion as an Energy Policy"? Unfortunately, it was because too few had the background to critically evaluate the president's statements. It's not that energy conservation is a difficult principle to understand. The subtleties arise because "energy" has many faces.

Energy is something that is easy to define but hard to comprehend. Energy is everything. The paper on which this book is printed is one form of energy. The light illuminating it is another, and the heat in the air where you now sit is yet another. The principle of energy conservation tells us that there is as much energy in the universe today as ten billion years ago; and there will be the same amount ten billion years in the future. The only thing that has changed, or will change, is the form this energy takes. As simple as this fact is to state, it took a long time to discover.

Our understandable preoccupation with kinetic energy explains why it took thousands of years before we learned to recognize energy in all of its forms. While one can see the motion of a thrown rock, and feel its energy if it strikes you, the same cannot be said of a stone perched atop a hill. Nevertheless, it, too, has energy, for with just a slight shove it will begin to move—which we call *potential energy*. A stone heated by a fire also has energy. That it has anything in common with the thrown stone or with the one perched atop the hill is anything but obvious.

By some accounts, a similar concern about apples in 1666 could have provided Sir Isaac Newton the necessary insight to link two of energy's forms. He knew that a falling apple possessed the most recognizable form of energy, that of motion. He also realized that hanging apples could fall and thus possessed potential for movement: potential energy. Applying the laws of motion that he had discovered, Newton could have determined exactly how much potential energy there is in an apple, or in any other object, including that stone on top of the hill.

On first consideration, there doesn't appear to be anything particularly special about either potential or kinetic energy. However, a simple experiment shows more is here than meets the eye. If the apple falls, it moves faster with each passing instant, transforming potential for motion into motion. Its kinetic energy increases while its potential energy decreases. Add the two together and we get a quantity with the uninspired name *total energy*. While the apple is falling, total energy is conserved; that is, it is constant: The apple has the same amount of total energy midfall as it does the instant before it hits the ground. This means that energy isn't being lost or created, it is just changing form. This is all that falling is—the transformation of energy from one form to another.*

Another aspect to this: When the apple hit the ground, however, its kinetic energy vanished without an increase in potential energy. Its total energy appeared to suddenly change and, as if by magic, energy disappeared. It required more than a hundred years to discover that energy *was* conserved and that the missing factor in the falling apple experiment was heat.

It would not be until the late nineteenth and early twentieth

---

*It is interesting that Newton never actually demonstrated that total energy was conserved for a falling object. Part of the reason was that he considered momentum as the key ingredient of *vis viva*, not energy. Newton's rival Wilhelm Leibniz was much closer to making the connection between *vis viva* and our modern-day interpretation of energy.

centuries that we would fully understand that heat is just another kind of energy of motion. Before this time, the prevailing view held that heat was something totally separate from energy—an invisible fluid, named "caloric" in the 1770s by Antoine Lavoisier, the father of modern chemistry. Artifacts of this stubborn and persistent notion are found in phrases like "heat flow" and "heat capacity," terms born of an unfortunate juxtaposition of technology with science.

In 1712 relating heat to energy became urgent, for it was then that Thomas Newcomen invented the steam engine. Originally designed to pump water from mines, the steam engine was nothing more than a machine for harnessing energy, which was just the brainwork we needed to get the Industrial Revolution under way. And this is why heat became important: The hotter a steam engine runs, the more work it can do. Throughout the remainder of the eighteenth century, people like James Watt and Richard Trevithick labored to improve steam engines by making them do even more work—and hence make more money.

As heat is typically the result of burning something, and burning was the province of the new science of chemistry, it seemed natural that chemists should be the people to explain heat. This was regrettable, for unlike physics, where we talk about mass, force, and velocity as concepts independent from the existence of real things, chemistry is about tangible elements and compounds. And so, it is not surprising that chemists should impress upon heat a physical form. In his seminal 1789 text *An Elementary Treatise on Chemistry,* Lavoisier begins his list of the thirty-three "known" elements with caloric, followed by light, then oxygen, nitrogen, and hydrogen. Two decades later, the Swedish chemist Jöns Jakob Berzelius classified caloric as one of five invisible and weightless substances, along with positive and negative electricity, light and magnetism.

Like its presumed vaporous counterpart static electricity, caloric was pictured as "self-repulsive," a property that was believed to explain why *everything* expands on warming. (This generalization is not entirely correct. Rubber, for example, contracts as temperature increases. Over a temperature range of a few degrees, so do water and plutonium.) Heating a thing was supposed to increase its concentration of this repulsive fluid, which should then force its constituents farther apart. While not particularly dramatic in solids, where the expansion is small, it is quite noticeable in a gas, like steam, where the volume change can be used to pump water or turn a gear. Frictional heating, for example, rubbing your hands together to warm them on a cold day, was neatly rationalized by assuming some caloric was "squeezed out" by contact, warming the surrounding area.

The flaw in the caloric theory of heat surfaced in a cannon factory in 1797. Count Rumford, born Benjamin Thompson, was the Commandant of Police at the court of Carl Theodor, Duke of Bavaria, and was responsible for the defense of Munich. Thompson had an early interest in heat, fire, and energy. As a thirteen-year-old living in his family home outside of Boston, he kept orderly notes for the construction of rockets and other fireworks. An unexpected explosion while assembling fireworks severely burned Thompson, but did not prevent him from pursuing his passion for heat. Thompson is credited with the invention of the double boiler, the kitchen range, and the Rumford stove. This latter invention brought him fame and fortune as it yielded more heat per pound of wood and eliminated smoke from the living space through an exhaust flue.

In 1775, having married a wealthy widow nineteen years his senior, Thompson settled in Concord (previously called Rumford), New Hampshire. A Tory, he spied for the royal authorities, passing notes written in another of his inventions, secret ink. He was arrested once, then released for lack of sufficient evidence. By 1776, the British presence in Boston was unsustainable and Thompson fled to Munich, abandoning his wife and their infant daughter.

The city the now-titled Count Rumford was charged with defending stood directly between the republican forces of the French Revolution and their enemy, the Hapsburgs of Austria. Though Bavaria was neutral in this conflict, Rumford ordered and oversaw the manufacture of heavy brass cannons to protect the city.

Cannons were first cast solid, and then the barrel was bored with a stationary hardened-steel drill bit held with great force against the cannon as it was rotated. The power for the process was provided by draft animals and transmitted to the cannon by a series of gears and pulleys. The frictional heating of the steel drill bit, in particular, would have been tremendous, causing it to glow ever so slightly in the dim light of the factory. This may have prompted Rumford to wonder how much heat—how much caloric—was in a cannon.

In no time at all, he set about measuring the heat liberated when metal rubbed upon metal. He cast a specially shaped insulated cannon barrel, replacing the sharp bit with a dull one, then immersed the whole thing in a tank of water to collect the heat released. As he wrote later, "I perceived, by putting my hand into the water and touching the outside of the cylinder, that Heat was generated; and it was not long before the water which surrounded the cylinder began to be sensibly warm." He goes on to write, "At two hours and thirty minutes it actually boiled. It would be difficult to describe the surprise and astonishment expressed in the countenances of the bystanders, on seeing so large a quantity of cold water heated, and made to boil, without any fire."

As might be expected of the inventor of the kitchen range, Rumford's imagination immediately turned to the practical: "[With] such a large quantity of Heat . . . produced . . . by the strength of a horse, without either fire, light, combustion, or chemical decomposition; and in the case of necessity, the Heat thus produced might be used in cooking victuals." Instantly tempered by scientific intuition, Rumford then notes that burning the horse's fodder might produce more heat and with far less bother, linking the drill's heat to the horse's digested oats and auguring the discovery of energy conservation fifty years later.

Toward the end of the experimental discussion, the count returns to his primary interest: "By meditating on [these] results, we are naturally brought to that great question which has so often been the subject of speculation among philosophers; namely: What is Heat? Is there anything that can with propriety be called *caloric*?"

He concludes in the negative, arguing the illogic that brass should contain an apparently unlimited amount of such a substance, lest the cannon barrel melt of its own accord. With characteristic brilliance, he reasons, "It is hardly necessary to add that anything which any *insulated body* . . . can continue to furnish *without limitation* cannot possibly be a *material substance;* and it appears to me to be extremely difficult, if not quite impossible, to form any distinct idea of anything capable of being excited and communicated in a manner that Heat was excited and communicated in these experiments, except it be *Motion.*"

Count Rumford's reasoning should have left the caloric theory in much the same state as its originator and chief proponent Lavoisier, who was guillotined in 1794 upon false accusations of corruption by fanatics of the French Revolution. But it did not. Rumford even refused to speculate on what it was that moved when heated. "I am very far from pretending to know how . . . that particular kind of motion in bodies which has been supposed to constitute heat is excited, continued, and propagated. . . . [And] I shall not presume to trouble the [reader] with mere conjecture." The failure to provide a mechanistic theory for heat and its transport doomed Rumford's theory to obscurity, where it would remain for more than twenty-five years until discovered anew, dusted off, and finally recognized as the foundation of the new science christened *thermodynamics*. A small triumph for the count: *thermo* comes from the Greek for "heat," and *dynamics* is the study of motion. Heat is movement and thus energy.

By the middle of the nineteenth century, many scientists knew of Rumford's cannon experiment and accepted his interpretation. Yet two questions remained. The first was the continuing concern over what exactly it was that moved in a heated object—the question Rumford wouldn't touch. Today we know that moving atoms and molecules "constitute" and "propagate" heat. But this realization came nearly one hundred years after the war for the atom was fought. For the chemists and physicists of the early nineteenth century, the battle lines had yet to be drawn. Atoms were, at best, heuristics, devoid of physical reality but useful for justifying the fact that elements combine in constant proportions. Connecting motion with heat added to the growing data, pointing to the existence of physical atoms. As the investigation of heat continued, the evidence would become overwhelming.

While the first question went unanswered, several scientists were determined to answer the second: How much movement produces how much heat? James Prescott Joule did it best.

Joule was a gifted and dedicated experimentalist, obsessed with finding the mechanical equivalent of heat. The supposed extent of his obsession is described in a story originally told by his friend William Thomson, Lord Kelvin. As the story goes, Thomson was vacationing near Mont Blanc on the French-Swiss border during the summer of 1847, when, during a walk, he chanced upon the honeymooning Joule and his new bride. Thomson was surprised to see Joule carrying a thermometer. On inquiring as to its purpose, Joule explained his intention to measure the temperature increase resulting from water descending a fall, a necessary increase if energy is conserved. So dedicated was Joule that he could not pass up an opportunity to work— even on his honeymoon.

The water at the bottom of a fall is indeed warmer than at the top, and the greater its height, the greater the temperature difference. But even in a very high fall, the temperature difference is a small one. So, apocryphal or not, the story captures the essence of the task confronting Joule: He had to determine the mechanical equivalent of heat by showing that a given amount of energy always produces the same amount of heat. And to do so, he had to make *very* precise measurements.

Joule was up to the challenge. With constant practice, he had learned to read a thermometer within $\frac{1}{200}$ of a degree Fahrenheit, a skill William Thomson called "magical." In addition, Joule employed two different techniques in order to bypass objections that results were subjective and dependent on details of the experiment.

In his first approach, Joule used electricity to produce heat. At the age of twenty-two, he had discovered that an operating electric circuit produces a predictable amount of heat, which depends only on the current and resistance of the circuit and the length of time it operates. In addition, he knew that a dynamo—a coil of wire that rotates in a magnetic field—could be driven by a slowly falling weight to produce a constant current. If the circuit were placed in an insulated container filled with water, the heat would raise the temperature of the water, which he could then measure with a thermometer. Joule reasoned that if energy were conserved, the potential energy lost by the falling weight would appear as heat, thus warming the water.

Joule found that a 772-pound weight slowly falling through one foot, or a one-pound weight slowly falling through 772 feet (or any combination in which the product of the weight with the distance fallen is equal to 772 (the units here are foot-pounds) raises the temperature of one pound of water by 1°F.

In his second experiment, Joule eliminated the dynamo (see figure). Instead of turning a coil of wire, the falling weight turned a submerged paddle wheel. After all, it didn't matter what forms energy went through as it turned into heat, it was conserved. As expected, he found the same relationship: 772 foot × pounds of energy provide the

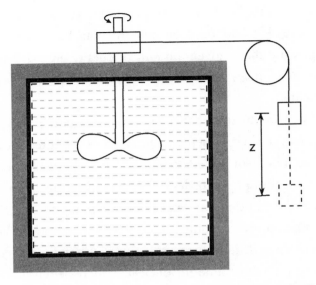

Figure 2.1 Joule's second apparatus employing a paddle wheel. In his first, the paddle wheel was replaced by a dynamo.

heat necessary to raise the temperature of one pound of water by 1°F. This amount of energy we now call the British thermal unit, or BTU.

This high cost—a lot of energy to produce a little heat—explains why the conservation of energy was not discovered sooner. The temperature of those apples falling from Newton's tree would have increased by a few hundredths of a degree Fahrenheit. If we had senses like pit vipers (rattlesnakes, copperheads), the conversion of kinetic energy to heat would be readily apparent. A pit viper "sees" heat and can discern tiny differences in temperature. To rattlers, an apple would visibly warm when hitting the ground. But alas, we have no such sense and had to wait for Joule and his magical skills with a thermometer to discover this most fundamental law of nature.

In recognition of Joule's accomplishment, the unit of energy in the metric system was named the *joule*. One joule is equal to roughly one thousandth of a BTU. The kinetic energy of a 2-kilogram mass

(4.4 lbs) traveling at one meter per second is equivalent to a joule. Another way to look at it: There are 1,463 kilojoules (1,000 joules) in a "good" candy bar (whereby I mean a loaded 350-calorie chunk of chocolate). Joule succeeded in illustrating that with the energy in such a bar, the body temperature of a 165-pound individual could be increased by approximately 8.4°F. Or that same energy could lift that same person 1.25 miles in the air, or propel him or her through a vacuum to a velocity of 440 mph. Since energy is conserved, this also works the other way around. To remove the energy of that single "good" candy bar, you would need to climb a 1.25-mile-high mountain, run very fast, or somehow raise your body temperature to dangerously high levels.

A few years following Joule's experiments with heat, the conservation of energy was enshrined as a basic principle of physics. Reformulated in mathematical form by the great theoretical physicist Hermann von Helmholtz, it became the First Law of Thermodynamics. There would be one more significant and totally unexpected addendum to the First Law in 1905, when, over a period of fifteen weeks, a young patent clerk published three papers that would revolutionize physics. That clerk, of course, was Albert Einstein.

Two of Einstein's three extraordinary papers provided insight into the First Law. The first confirmed that atoms and molecules communicate and propagate heat. The confirmation came as Einstein explained why smoke particles in air exhibited random and jerky motion. He showed this to be the result of collisions between the smoke and much smaller moving particles. Though it came hundreds of years too late for Count Rumford, this was the final piece of evidence needed to convince many scientists of the existence of atoms and molecules. Now the full story of heat was clear. Heat was nothing more than the total kinetic energy for all the molecules in a system, say a cup of water or a bowl of soup. Temperature is not the same as heat; temperature was seen to be related to the average kinetic energy of the molecules in a system. Though a big system may be at a lower temperature than a small one, the big system—containing a great many molecules moving slowly—could contain more heat than a small system where the molecules were moving very fast.

The story of heat was clear, but no one had anticipated the conclusions that would be drawn from Einstein's second paper. Einstein indicated that mass was just another form of potential energy. The famous equation $E = mC^2$ (energy is equal to mass times the speed of light squared) shows that just as heat is equivalent to energy, so, too, is mass.* With this expanded insight, the full statement of the First Law requires that mass/energy be conserved. The universe is composed of nothing but energy packed together in various ways; everything that we observe involves energy converting between its several forms—kinetic, potential, heat, or mass—while its total amount remains the same.

Based on what we know now of the First Law, recall that President Bush promoted the development of a hydrogen fuel cell vehicle that would "reduce our dependence on foreign oil." The logic goes something like this: Hydrogen is abundant. We find it everywhere, including in water. When hydrogen combines with oxygen in fuel cells, energy is produced. The energy can be used to propel cars or other vehicles, and the neat part is that the sole by-product of this reaction is water. We start with water, run our cars, and get water back. What could be better? Nothing, as long as you ignore the fact that the energy to run our cars has to come from somewhere. We started with water, ended with water, and got motion in the bargain. The flaw is that the fuel to run our cars is not water but pure hydrogen, which must be extracted from the water. And guess what? This requires energy. Energy is used to make hydrogen from water. This energy is recovered when the hydrogen is turned back into water and used to run our cars.

---

*This means that the total energy content of a one-kilogram mass is equivalent to that of more than 61 billion good candy bars.

Then energy is conserved, balance is restored, and all is right with the world.

Where was Bush planning to get the energy needed to extract hydrogen from water? By burning fossil fuels. Our president was waving a magic wand; he pulled a sleight of hand. Not as obvious as making things disappear, but with a little misdirection, Bush tried to circumvent the First Law.

Our first lesson in diet basics has been that energy is conserved. There will be no free lunch. The energy to run our cars and power all of our electrical appliances must come from somewhere. As attractive as an idea may look, we must ask, Where is the energy coming from? Where is the energy coming from to make hydrogen? Where is the energy coming from to make ethanol from corn? This is only the first question that must be asked.

Because heat is energy, there is energy coming out of the tailpipe of your car. Hence it seems perfectly reasonable that we should be able to harvest this energy and feed it back into the engine. We aren't creating energy just by using it again and again. However, although energy is not consumed, it does become "less nutritious." The energy being expelled through the exhaust system is less nutritious than that contained in the gasoline injected into the car's engine.

This brings us to the Second Law of Thermodynamics and the second question we will ask when constructing The Thinking Man's Energy Diet: How much nutrition is in that energy?

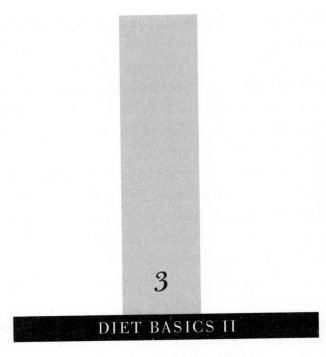

3

All Energy Is Not Equal

The second George Bush was not the first U.S. president to play fast and loose with the laws of nature. That honor may belong to President James A. Garfield, who, in the 1880s, enthusiastically supported the development of an engine that ran on ammonia and its vapor, in place of water and steam.

John Gamgee, an engineer trained in England who was working in Washington during the Garfield administration, assumed that the flaw of the steam engine was in the tremendous amount of energy it required to boil water. Gamgee reasoned that an engine using a fluid with a lower boiling point, like ammonia, which boils at zero degrees

Celsius, would be more efficient—from which Gamgee's creation, the *zeromotor*, takes its name.

The idea behind the zeromotor was to boil liquid ammonia using the energy of seawater. Seawater would be circulated in a coil through liquid ammonia and the heat from the seawater would boil the ammonia and vaporize it. The expanding ammonia gas could then be used to drive a piston, and then be recycled trough the traditional condenser found in all steam engines. There it would be turned back into a liquid, ready for the next cycle. As long as the temperature of the water was above freezing (the boiling point of ammonia), no fuel should be required. If the zeromotor worked, its power would come from heat energy extracted from seawater.

At the time, the U.S. Navy was dependent on a costly and extensive network of coaling stations strategically located throughout the world. The possibility of powering ships with heat extracted from the very waters on which the vessels sailed was more than the navy's chief engineer, B. F. Isherwood, could resist. He recommended further development of the ammonia engine to U.S. cabinet members and President Garfield, all for a working model that, in the end, functioned only with the condenser disconnected—and then poorly.

That the zeromotor worked at all indicates it did not violate the law of energy conservation. Real joules, extracted from seawater or the air, were used to power this engine. What made it run so poorly was the *quality* of the energy. These were not very nutritious joules. It is the Second Law of Thermodynamics that encompasses the quality of energy—a law that should have been well-known to Isherwood.

Because the universe is filled with many different forms of energy (and only energy), everything that happens is a manifestation of energy changing form: from mass to heat, from kinetic to mass, from potential to kinetic, etc. And in every instance, total energy is conserved. If X number of joules exist before something happens, X number of joules must exist after the event has occurred. But this is all that the First Law tells us. As far as it is concerned, potential energy may be converted to heat as readily as heat is changed to potential energy.

The Second Law, however, is about nutrition; it tells us that when it comes to doing useful things with energy—work—all forms of energy are not equal. Work is the form of energy that lifts weights, moves objects, or powers a ship. The part of energy that is less utilitarian, that is, composed of "empty calories," as my mother would have said, is called heat.

We've established that everything happening around us involves energy changing form. In doing so, some energy does work and the remainder becomes heat. The Second Law controls the amount of energy that becomes heat. It also constrains the amount of work that can be realized from heat. One statement of the Second Law (there are many and they are all equivalent) forbids heat to be entirely changed to work. Kinetic energy can be changed fully into work, as can potential energy, but not heat.

Remember Newton's apple? It fell, and in the process converted potential energy to motion and then to heat. The apple cannot now take that heat and transform it totally into potential energy to leap back into the tree—not because it would violate the law of gravity, but because it violates the Second Law of Thermodynamics. Thus, it is the Second Law that separates what is possible from what is not, and it also separates the past from the future. Though the connection between the generation of heat and the flow of time is not readily apparent, these phenomena are intimately intertwined in the Second Law of Thermodynamics. In many respects, this law is nature's most mysterious and least understood.

If you are having trouble with these ideas, bear with me. It took scientists two hundred years before they began to understand the ramifications of the Second Law of Thermodynamics. And it is not clear that the full picture has yet emerged. So take a deep breath, and give me three chapters to see if I can get you to the point where all of this makes sense.

Even before the First Law had been fully elucidated, the French military engineer Sadi Carnot (1796–1832) was hot on the trail of the second. Carnot, a fervent nationalist, had undertaken an intense study of industrial development, with the objective of enhancing French influence in the world. Eventually Carnot came to believe that true power and advancement was to be found in technology, particularly in the steam engine. He attributed Great Britain's dominance to the English and Scottish engineers who had labored for more than fifty years to improve steam-engine efficiency. All the while, the French contributed virtually nothing. Writer Hans Christian Von Baeyer notes that so important was the steam engine to England's industrial strength that Carnot conjectured, "The destruction of [England's] Navy, which she considers her strongest defense, would perhaps be less fatal [than the loss of the steam engine]."

Engineers of the time, like the Scot James Watt, had taken a pragmatic approach. Their goal was to better steam-engine efficiency, which required increasing the work that could be done per ton of fuel burned. They tinkered with design and materials, sought out and plugged sources of heat loss, and eliminated vibrations, resulting in efficiencies of about 6 percent. In other words, for every 100 joules of energy produced by burning wood or coal, only 6 joules were used to move or lift something; the leftover 94 joules remained as heat and were dumped into the environment.

It never occurred to the designers to ask, "What is the theoretical maximum efficiency of a steam engine? Is it 6 percent, 20 percent, or even 100 percent?" Until then, the engineers had no idea if they were making minor improvements or if they were already pushing the limits of steam-engine design. Carnot was the first to ask this obvious question.

Named for a medieval Persian poet, Sadi Carnot was born in Paris in 1796. After obtaining a solid education in mathematics, science, language, and music, he studied engineering under the direction of his famous father, Lazar Carnot, a general and Napoleon's minister of war. Sadi joined the military for a few years before retiring as a lieutenant and moving back to Paris at the age of twenty-four. There, with only the pension he received from the army as support, he turned his full attentions to the analysis of the steam engine.

Carnot approached the problem as a scientist, seeking to develop a general theory to explain how heat can be converted to work. (Remember, the Second Law forbids heat's entire conversion to work, not its partial conversion.) Just as Newton had developed a theory to explain motion that applied equally to planets, cannonballs, and apples, Carnot sought a theory to explain the motive power of heat that was independent of the machine performing the conversion. As he would later state in his *Reflections on the Motive Power of Fire:*

> In order to consider in the most general way the principle of the production of motion by heat, it must be considered independently of any mechanism or any particular agent. It is necessary to establish principles applicable not only to steam engines but to all imaginable heat-engines, whatever the working substance and whatever the method by which it operates.

Though his aspirations were both appropriate and eminently feasible, Carnot was saddled with a serious handicap. He was working at that inopportune time when the caloric theory of Lavoisier still held sway, particularly in France. So Carnot's conclusions are based on two axioms: the impossibility of perpetual motion and the fluid theory of heat. The true testimony to Carnot's genius is that he came to correct conclusions from flawed principles.

What Carnot realized, which had escaped all others, is that heat engines, regardless of the design or the materials from which they are made, dump heat into the surrounding environment. Steam engines expel heat up the smokestack. (The same goes for the heat engines of today: Automobiles dump heat through the tailpipe and radiator, and nuclear reactors leak heat into cooling towers and then to the atmosphere.) As hard as the engineers of Carnot's time tried, they could not eliminate this massive heat loss.

Carnot guessed, correctly, that this heat loss was a necessary consequence of harnessing the power of heat. He pictured heat in a fluid—caloric—form and drew analogies between the flow of heat and the flow of water through a waterwheel. Just as water flows downhill and can turn a wheel, he reasoned that heat flows from hot to cold and can do work as it proceeds. Thus, the flow of heat to the environment is as much a part of extracting work from heat as the flow of water is through a waterwheel. After having convinced himself of this fact in 1823, Carnot declared a general law: It is impossible to extract work from heat without simultaneously discarding some heat. This would become yet another statement of the Second Law of Thermodynamics.

Though Carnot's conclusions about the Second Law were correct, his rationale, based on the caloric theory of heat, was wrong because according to that theory caloric was a physical substance and must be conserved, just as flowing water is. With heat conserved, the amount of heat expelled to the surroundings would be the same as the heat produced by the burning of the fuel. The energy going into the steam engine (in the form of fuel) is equivalent to the amount coming out in heat, but as we now know, some energy is used by the steam engine to lift or move something. Thus more energy comes out than went in—a clear violation of the First Law. Unfortunately for Carnot, in his time the First Law was unknown.

The conceptual flaw in Carnot's reasoning did not invalidate his statement of the Second Law. With just the assumption that perpetual motion machines were impossible, Carnot could have come to the same conclusion without reference to caloric theory. Consider a machine that could extract work from heat without waste. With this apparatus, we could build a completely efficient electrically powered machine that pumped heat from a box and made it colder. Sound familiar? I just described a refrigerator. We could then recover this heat, convert it to electricity, and run the machine. Together, the refrigerator and heat recovery system would form a perpetual cooling machine, requiring no external power whatsoever. Heat would flow naturally from a cold object to a hotter one. Because we began with the assumption (based on millennia of observation) that such machines cannot be, we must conclude that it is impossible to create a machine or a process that converts all heat into work, as Carnot realized. As a bonus, we arrive at an alternative statement of the Second Law: It is impossible for heat to flow from cold to hot.

There is evidence in Carnot's notes that he struggled with the paradox of heat conservation as required by the caloric theory. Toward the end of his tragically short life—he died of cholera at age thirty-six—it appears that he was about to abandon this theory and make estimates of the mechanical equivalent of heat, as Joule did. Had he done so, he would have realized that some heat had been lost as work. Perhaps, had he survived longer, he would have made this leap and become the true Father of Thermodynamics. Adding the discovery of the First and Second Laws to his name, his achievements would have rivaled those of Newton, Maxwell, and Einstein.

Despite his conceptually flawed foundation, Carnot went on to develop a solid framework from which to analyze the efficiency of a heat engine, a framework that survived even after it was accepted that energy—not heat—is conserved. Fundamental to this framework is the concept of a reversible machine, one in which the only heat loss is that required by the Second Law.

Up until Carnot, no one had imagined that some heat loss was actually necessary. In order to isolate only "wasted heat," Carnot envisioned a machine that moved infinitely slowly. It would begin in one state and, at an infinite time later, return to that same state. As an example, let's imagine a cylinder with a piston that can support varying weights. As weight is added, the piston moves, compressing the gas in the cylinder. Now take the whole apparatus, place it in a watertight container, and put this in a bath of water. Any heat generated by the motion of the piston will warm (or cool) the water, which can be measured by monitoring its temperature. Now we will do two experiments with the apparatus.

In our first thought experiment, a one-kilogram weight is placed on the piston and then released. The piston falls, rebounds, then falls again. These oscillations continue until the piston comes to rest, where the pressure in the cylinder *just* balances the applied weight. The process is then reversed and the one-kilogram weight is removed. Again, the piston bounces up and down and comes to rest near, but not exactly at, its original position. The temperature of the water surrounding the experiment is found to be higher than at the start, indicating some of the energy used to move the piston was converted to heat.

For our second thought experiment, we will add the weight slowly by using one kilogram of talcum powder. Very gently, the patient scientist will use forceps to place single grains of talc on the piston until the total weight reaches one kilogram. Presumably, the small oscillations that result from the placement of each grain of talc are imperceptible. Next, the talc is removed, again one grain at a time, until the piston no longer supports a load. Now the position of the piston will be very, very close to its starting point, and the temperature of the water bath should be undetectably warmer than at the start. On the off-chance that we can measure a temperature difference, the experiment can be repeated with progressively finer grades of talc until the limit of our ability to measure temperature differences is reached.

The second thought experiment is reversible. The piston and water bath can be restored to their starting state, and in so doing no net energy is converted to heat. Any process that can be returned to its starting condition without an energy cost is called reversible (whether or not the process is actually returned to its starting condition doesn't matter). While a reversible process making a full cycle converts no energy to heat, during the reversible compression phase in our second experiment heat is absorbed from the cylinder by the water bath; during the reversible expansion phase this same amount of heat flows from the bath into the cylinder. These heat flows are necessary. The first thought experiment is irreversible, however, because there is no way to get back to the starting condition without expending energy.* An irreversible process will always generate more heat than a reversible one. Herein lies the opportunity to reduce unnecessary heat loss through design.

Using the concept of reversibility, Carnot formulated his most important theory. Recall that Carnot envisioned a heat engine as working between two thermal reservoirs, one hot and one cold. The transfer of heat between these reservoirs provided the means by which heat produces motion. Through sheer logic, Carnot proved that the efficiency of a reversible engine depends only on the temperatures of the two heat reservoirs between which it operates—the greater this difference, the more efficient the engine. To put it another way, the nutritional content of heat is determined by the temperature difference between two ther-

---

*Another way to test for reversibility is by imagining a movie of the process. A reversible process would look reasonable, even if the movie were being shown in reverse, whereas an irreversible process would not.

mal reservoirs. Only if the difference is truly huge (infinite), will the heat be 100 percent "nutritious."

This brings us back to the flaw in Gamgee's zeromotor. The temperature difference between ammonia and its vapor is small—much smaller than the temperature difference between the steam in the boiler and the water vapor in the condenser of a steam engine. Thus, the zeromotor could not possibly operate more efficiently than a steam engine, and naval engineer Isherwood should have known better.

So energy is conserved, and heat must flow from hot to cold. While the First Law sounds somewhat "scientific," as a deep fundamental truth of nature should, the Second Law seems almost commonsensical. This impression derives from the scientist's bias for quantification: The First Law translates easily into the preferred mathematical language.

It is like the money in your bank account. Add up the desposits, subtract withdrawals, and that is the amount you've got. But the Second Law of Thermodynamics did not seem to lend itself to this tidy form of bookkeeping. As a pair, the Laws of Thermodynamics appear to be an odd couple.

At least one scientist sought a more appropriate statement of the Second Law. Steeped in the German scientific traditions of his time, Rudolf Clausius was offended by the asymmetry between the first and second laws. Though his 1850 paper was instrumental in rehabilitating Carnot's theories, after the First Law had been firmly established he did not rest until the Second Law was placed on equal footing. His innate persistence produced results with the 1854 publication of a paper titled "On a Different Form of the Second Law of Thermodynamics." What Clausius had discovered was a previously unknown quantity that was conserved in all reversible processes. He called this quantity *entropy,* from the Greek word for "transformation," explaining:

I have specifically [chosen] the word entropy so as to be as similar as possible to the word energy, since both these quantities, which are to be known by these names, are so nearly related to each other in their physical significance that a certain similarity in their names seemed to me advantageous.

What is entropy, exactly? Before we can genuinely appreciate Clausius's wonderful insight, we need to get comfortable with the lingo of thermodynamics.

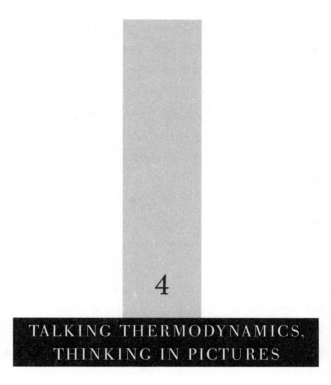

4

TALKING THERMODYNAMICS, THINKING IN PICTURES

Entropy

Much as Eliza Doolittle was discounted before she learned to talk the talk, scientists are just as dismissive of those who don't express themselves correctly. Eliza spent months learning to enunciate, walk elegantly, and dance, but nonscientists need only learn a few words and repeat a few mantras if they wish to be inconspicuous among a school (perhaps *college* is the correct word) of scientists. Science is all about measurement, and these words and mantras deal with making and interpreting the results of measurements.

At any one time, measurements can only be made on a small piece of the universe, hence we distinguish that which we are measuring as "the system" and everything else as "the surroundings." Implicitly, there is a boundary between the system and its surroundings, and though that boundary may correspond to something real—say, the surface of a test tube—it need not be anything real at all. In actuality, it may exist only in the mind of the experimenter.

One of the advantages to partitioning the universe into a system and its surroundings is the ability it gives us to measure conserved quantities. For example, energy is a conserved quantity, and we know that if we consume 3,000 calories in a day, and between exercise and our resting metabolism we have converted 2,500 of those calories to heat, then 500 calories remain in our body (the system) and will become fat. We did not have to wait, or perform some invasive test, to find out how many calories were hiding out inside us; we knew the quantity because energy is conserved. What we didn't convert to heat is still inside the system. Consider now a balloon filled with air at a fixed temperature. We will define the region inside the balloon as the system, the region outside as the surroundings, and the balloon itself as the boundary between the two. If the balloon is now cooled, causing some quantifiable amount of energy to be lost from the system and gained by the surroundings, we can picture this energy flow as taking place across the system boundary. This image has heuristic properties that make it very useful to scientists, for now we need only make measurements on the system to determine how much energy (or any conserved quantity, for that matter) has been lost or gained. Just imagine how complicated it would be to make the same measurement if we had to "find" the energy lost by the system in the surroundings. We would need a huge number of sensors—thermometers, pressure transducers, and such—and we would always be questioning whether or not enough of the surroundings had been included in our measurements.

Next, we will need to describe the "state" of our system to other scientists. Although I can do an experiment in Colorado, unless someone can repeat that experiment in the Netherlands or Hong Kong and get the same answer, it is of little use: Science demands repeatability and consistency. So, I want to describe my experiment to other scientists who will attempt to repeat it and, I hope, get the same results. I could give a full description, in which a yellow pear-shaped balloon containing one gram of air, pressurized to 1.1 atmospheres (an atmosphere is the pressure produced by the air at sea level, about 15 lbs per square inch), at 22°C, was cooled to 0°C in a large aluminum baking pan containing 500 grams of ice and 500 grams of water. Not surprisingly, anyone who repeated the experiment using these instructions would see the same amount of energy flow from the system to the surroundings. But so would an experimenter who used a red oval-shaped balloon or an experimenter who cooled the system in a bathtub of ice water. In fact, anybody who cooled one gram of air at 1.1 atmospheres from 22° to 0°C would get the same result. Thus we say the state of our experimental system is specified by four quantities, or *state variables:* the substance (air), its mass (one gram), pressure (1.1 atmospheres), and temperature (22° or 0°C). It is only necessary to know the values of these four variables in order to reproduce the experiment.

Rarely, if ever, are we presented with a unique set of state variables. For example, we could have specified the volume of the balloon, omitted the temperature, and still have determined the amount of energy transferred from the system to the surroundings. This is because temperature depends on the amount, volume, and pressure of the air. For a given amount of air at a given pressure and volume, the temperature is predetermined. Therefore, temperature, volume, and pressure are all state variables, but we do not need to know all of these to specify the state of the system. We need specify the values of only three of the four. The three we choose are called *independent variables,* the remaining one, a *dependent variable.* The choice as to which variable will be dependent is arbitrary and generally picked to simplify an experiment; for example, it is easier to measure volume than pressure, so often we treat pressure as the dependent variable.

Consider now another system, one characterized by a set of independent state variables that we will monitor. As long as these variables are unchanging, then the system is doing nothing (at least at the macro level). For all intents and purposes, it's just sitting there. We can activate the system, so to speak, by doing something that causes some or all of the state variables to change—heat the system, squeeze it, shake it, spin it. The result is a *process*.

As an illustration, let's use another balloon containing one gram of air. The state of this balloon can be characterized by two independent state variables (remember, the amount of air in the balloon is fixed and thus cannot vary). Let's pick pressure and volume, and we'll indicate these with a pair of numbers in braces: $\{P, V\}$, where P is the pressure and V is the volume. Now we can visualize the entire process by reducing it to a simple line drawing on a grid (by convention the first dependent variable, P, lies along the horizontal axis and the second, V, along the vertical). Now, start with a balloon subject to 1 atmospheres of pressure (about 15 pounds per square inch) and with a volume of three liters: $\{1, 3\}$. Then cause it to change to a balloon subject to three atmospheres of pressure and a volume of half a liter: $\{3, 0.5\}$. (The temperature will also change, but we can always find the temperature by knowing the pressure and volume. In this case, temperature is a dependent state variable.) We represent this change with a graph, shown in the following figure.

The figure depicts the starting and ending state of the balloon, as well as three processes connecting its initial and final states. Each of the processes shown is reversible, because we could make the balloon go back along the path traced. However, to move along these reversible paths requires the balloon to change infinitely slowly. There are also irreversible processes that can't be shown on the graph. To make this point clearer, consider what would happen if you rammed the balloon into a half-liter container. A shock wave would move through the bal-

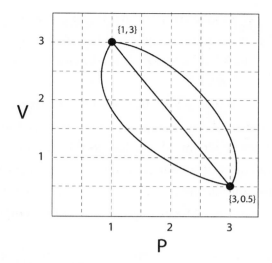

Figure 4.1 Reversible processes for a balloon

loon, bounce off the far side, and return. If you measured the pressure in the balloon, it would be different at different points and changing in time. There is simply no single pressure, and hence no point on the graph, that corresponds to the state of the balloon. You would have to wait some time before all of the state variables stopped changing.

Clearly, an infinite number of processes take the balloon from its initial to final states. The question that naturally presents itself is: What are the differences? And that brings us right back to Rudolf Clausius's entropy. What he found was that though two processes may start and end at the same point, the amount of energy converted to work and heat is not the same. He discovered how to put a number to the efficiency of a process, where the more efficient process produces the most work and least heat. And this is where entropy comes into the picture.

Clausius discovered that in going between two states, a reversible path always produced the least heat and the most work. He showed that if we call the amount of heat produced along one of these reversible paths Q and the temperature T,* then the entropy of the system will change by an amount equal to Q/T. We call the change in entropy of the system ΔS_{sys}, where the Greek symbol Δ is used to denote "change."

This is all well and good, but what about irreversible changes between two states? Then, more heat than is required to move between these states is generated and lost to the surroundings; let's call this Q_{sur}. Thus there is a change in the entropy of the surrounding, which is given symbolically as $\Delta S_{sur} = Q_{sur}/T_{sur}$ (where the subscript *sur* obviously denotes surrounding, so T_{sur} is the temperature of the surroundings).

Now here is the key part. If we add the change in entropy of the system to its change in the surroundings we get the entropy change for the universe as a whole; we will call it ΔS_{univ}. What Clausius found was that $\Delta S_{sys} + \Delta S_{sur} = \Delta S_{univ} > 0$ for all processes that happen in finite time, i.e., all irreversible processes.

Of course, these symbols are exactly what Clausius was looking for—a nice mathematical statement of the Second Law of Thermodynamics—but don't let this intimidate you. To truly appreciate what Clausius found, you must see more than symbols and equations. These are but the threads of a magnificent tapestry. The equation $\Delta S_{univ} > 0$ speaks of the ultimate irony and tragedy of nature—this law gives rise to existence and at the same time guarantees that it is fleeting. You need not be a scientist or mathematician to grasp this fact; you only need to be trained to see past the symbols to the story underneath.

To find the encrypted story, begin by forming a picture in your mind's eye. What image do you summon up for the symbols Q and T?

---

*Temperature must be measured relative to absolute zero. This was discovered by Joule's friend William Thomson, Lord Kelvin, from which the absolute temperature scale takes its name. Zero Kelvin, 0 K, is equal to −273.15°C or −459.67°F.

My image of heat involves a container filled with moving balls, colliding and rebounding in a never-ending dance of confusion. The *total* kinetic energy for these moving particles is heat, which means that a large number of particles moving slowly may contain just as much heat as a smaller number moving much faster. What distinguishes these two systems is their temperature, which is related to the *average* kinetic energy of the balls. Remember, kinetic energy is related to velocity. Thus the slowly moving balls are at a lower temperature than the faster ones.

Now turn the whole thing into something familiar, say Kool-Aid. To prepare a drink, the instructions suggest mixing a package of colored crystals with two quarts of water, yielding a beverage with an *average* concentration of one-half package per quart. Add more water and the result is weaker and less flavorful; add too little water and the beverage is too sweet.

We can draw an analogy between heat, temperature, and entropy and the time-tested recipe for Kool-Aid. Make the *total* number of packages used to make the drink analogous to heat, and then the strength of the drink—its concentration (the *average* number of packages per quart)—is analogous to temperature.

Let's say we find four quarts of cherry Kool-Aid in the refrigerator. Assuming it was properly prepared, it contains two packages of Kool-Aid crystals. The concentration of the drink is ½ package per quart, so a quantity analogous to entropy—heat divided by temperature—can be found by dividing two packages by ½ package per quart, which gives us four quarts. Result: Entropy is analogous to the amount of liquid used to dilute the Kool-Aid crystals. Remember the part of the Second Law that requires that the change in entropy accompanying any nonreversible process be greater than zero? This is tantamount to saying that the only thing we can do to our Kool-Aid is to add water and make it more diluted. In essence, the Second Law requires energy to spread out and become more dilute over time.

There is only a finite amount of this substance called energy. The transformation between its forms is what makes things happen. Yet with every transformation the energy becomes a little more diluted. In the distant future, energy will become so thin, so dispersed, that it will no longer provide the "nutrition" necessary to keep our universe going and the cosmos will settle into a long hibernation from which it may never wake. This is called "heat death." From what we know now, there does not appear to be a way around the ultimate heat death of the universe.

For a better grasp of the steps leading to the universe's heat death, think of a mundane heat transfer, such as a teacup full of scalding hot water sitting on a coffee table. As soon as the cup is filled with water, it begins to cool. We'll now consider the entropy change associated with the transfer of that first joule of energy. For such a small amount of heat, the temperature of the recently boiling water will hardly change; hence we can treat it as constant, 373 K (212°F), and the entropy lost from the cup will be $-1/373$, or approximately -0.00268 entropy units.* The joule lost from the cup will flow into the air at a temperature of 300 K (81°F). The entropy gain of the air is therefore $1/300$, or approximately 0.00333 entropy units. The total entropy change, cup plus air, is $0.00333 - 0.00268 = 0.000653 > 0$. The Second Law of Thermodynamics permits this irreversible process; that is, heat flowing from hot to cold. More important, irreversible processes amount to a dilution of energy, which is the lesson here. The concentrated heat energy in the teacup is being diffused into the air.

We know heat transfer will stop when the cup of tea reaches the temperature of the air: 300 K. But instead of considering temperature, think of the process in terms of entropy change. The last joule of energy transferred to the room, before the cup and air come to thermal equilibrium, will remove $1/300$ entropy units from the former hot water. In turn, the room gains $1/300$ entropy units, yielding a total entropy

*An entropy unit is a joule per degree Kelvin, as in J/K.

change of zero. The cup and air are now undergoing reversible heat transfer. Conceptually, the energy concentrations in the room and the cup are equal.

In a reversible process, entropy is conserved; in an irreversible process, it increases. Is there a process where net entropy decreases? It is easy to construct the right set of circumstances by simply reversing the situation above. Assume the air transfers one joule of heat to the boiling water. Then the entropy of the air changes by −0.00333 entropy units and the cup gains 0.00268 units. The total change is −0.000653 entropy units, which the Second Law forbids. This is neither a reversible nor irreversible process. It is an impossible one. The Second Law reflects the fact that we never see heat concentrate in an object at the expense of its surroundings. No matter how long we wait, a warm room won't heat a cooling cup of tea and weak Kool-Aid will not sweeten of its own accord. The conclusion is crystal clear: Processes that lead to a net entropy decrease are, by all observations, impossible.

Clausius understood the Second Law of Thermodynamics to mean that for any spontaneous process (by which we mean something that *can* happen, whether it does or not) the accompanying entropy change must be greater than or equal to zero. The first and second laws now take on a similar form that could be expressed symbolically as $\Delta E_{univ} = 0$ and $\Delta S_{univ} > 0$, respectively. In words, these simple mathematical expressions tell us that for every process, the exact amount of energy lost (gained) by the system must be gained (lost) by the surroundings. Some of this energy will take the form of heat, so that the entropy of the system plus the surroundings is greater at the end of the process than when it started. Putting the two laws together, we know that as situations occur, the energy in the universe is being diluted. Picture it:

During a game of pool, a player breaks and the cue ball dilutes its energy across the other fifteen balls on the table. In turn, these roll to a stop after diluting their energy on the table's felt covering. Athletes dilute the potential energy of food in sweat. The concentrated energy of gasoline is transformed into hot brakes and cloudy exhaust. Everywhere we look, the reality of the Second Law insinuates itself.

The Second Law also differentiates the past from the present. A billion years ago, a million years ago, or just an instant ago, energy was more concentrated than it is now. In other words, the past is measurably different from the present or the future. All other laws of nature are apparently oblivious to what seems so obvious: Time moves forward. Returning to Newton's apple, the reason it fell to the ground and stayed there is not because of gravity per se; some of a falling apple's energy is converted to heat, which, in turn, generates entropy. Hence, the direction in which entropy increases corresponds to a falling apple. The mathematical equations that govern the motion of apples do not preclude the apples from leaping off the ground and into the trees. However, if I were to show you a movie of an apple doing just that, you would know that it was being run in reverse. In this direction, entropy decreases and energy becomes more concentrated.

What would happen if the Second Law were suspended? Would apples be able to leap back to a tree's branches?

If the Second Law were suddenly null and void, then a falling apple would generate no heat. There would be no friction as it fell through the air, and it would not warm the ground or itself upon impact. Instead, its energy of motion, which still must be conserved, would be consumed as the apple *elastically deformed* upon hitting the ground. And, like no ball you have ever seen, the amazing fruit would rebound from the ground to the exact height from which it fell—and then fall again. It would bounce forever between the ground and the branch on which it ripened. There would be no clue as to whether a movie of the jumping apple was being shown forward or in reverse, for with the temporary suspension of the Second Law, time loses direction.

How is it that a law requiring heat to flow from hot to cold can also give rise to time? Somehow, nature was able to produce a complex and diverse universe from a simple set of laws. Of these, it's the Second Law that stands out as unique. In fact, it really isn't a law at all, it is a suggestion.

Energy conservation is a "real" law of nature. No matter how hard you try, it cannot be broken. The same goes for all conserved quantities, such as momentum, charge, spin, etc. Luckily, the Second Law does not strictly concern itself with conservation. If it did, and entropy was conserved, the universe would never change. The beginning, middle, and end of time would look the same. People would not come into being and would not ultimately die. There would be no art, literature, music, or science. There would be no brainwork of any kind, since brains make heat and generate entropy.

There would also be no freedom of choice, because it is the Second Law that allows us to choose. Of the many different choices we make, the lasting consequence of each is written in entropy. Years after paying the ticket for running that red light, or taking the elevator instead of walking up the steps, the universe will carry the entropy burden generated through those choices. Billions of years from now, the entropy you bequeathed the cosmos during your life will still be around. It is the Second Law that lets you choose the size of this bequest. The only thing this law forbids is do-overs. You only have one chance when it comes to your entropy legacy.

So, you can elect to produce entropy judiciously or not, what difference does it make? The universe isn't going to "punish" you for *foolhardy* entropy decisions. Or will it? It will and does.

In support of this claim, consider the simplistic popularization of the theory of natural selection as "survival of the fittest." (Note that the theory of natural selection is not the same as the theory of evolution. On the contrary, it is the mechanism through which evolution acts.) The terminology *survival of the fittest* is not Darwin's; nonetheless, it has become intimately associated with his theory. In this form it has served to justify some of the darkest episodes of our history. The problem is its circular nature—one can argue that survival implies fitness. The Nazis interpreted natural selection in this way, rationalizing the systematic extermination of Jews. As they saw it, the failure of the Jews and other persecuted minorities to survive was attributed to a lack of fitness, and the survival of the German people proved their intrinsic superiority.

To expose this perverted misuse of reputable science, a concrete definition of *fitness* is required. Take Lance Armstrong as an example of someone who is fit, compared to me. How would we go about quantifying our relative fitness? One way is to measure the energy used to do the same amount of work. Let's imagine that Lance is participating in this experiment. First, he and I use weight belts to equilibrate our weights. Then we hop on identical bikes and begin a ride up a spectacular mountain pass in the French Alps. As we ascend, food energy is converted into potential energy and into the heat radiating from our bodies, which we measure with heat sensors. Lance and I generate the same potential energy (we are lifting equivalent weight to the same altitude), but the heat produced is quite different: I generate between 40 and 100 percent more than he does. Quite simply, Lance is more efficient than I am, getting more work and producing less heat from food energy. This is why he is more fit. In other words, he generates less entropy than I do for the same bicycle ride.

"Survival of the thermodynamically more efficient" may be a clumsy phrase, but it is not circular or prone to misinterpretation. I am proposing we use thermodynamic efficiency as a measure of fitness. This quantity is arrived at by the percent of energy input that results in work; the remainder shows up as heat and therefore entropy. With this definition in hand, come to the African savannah with me, where we'll see two cheetahs fighting to survive. One of these speedy cats is more efficient than the other. This means that it wastes less energy when hunting and therefore can afford to expend more in the chase. With more energy going to work and less to heat, it runs faster and has a higher hunting success. If prey should become scarce, it is the more efficient cheetah that is likely to survive. Nature rewards efficiency and punishes the producers of excess entropy.

As we have seen, the universe is struggling to survive, struggling to stave off heat death. Maybe it is not competing with other universes (or maybe it is), but it is still fighting for a long life. Toward this end, nature produces entropy just fast enough to move things along but not so fast as to end existence prematurely. In this respect, all of existence is like a grandfather clock, where energy is initially stored in elevated brass weights. By design, these weights are not allowed to fall freely and convert their potential energy into heat; rather, over a period of days, the stored energy becomes heat as the weights drop slowly, moving a pendulum, gears, clock hands, moon dials, and ringing chimes. Whether the weights plummet in a flash or make a controlled descent over the course of a week, the same amount of energy is converted to heat. The rate of entropy production is slower when the clock operates as it is designed to do.

Through myriad examples, nature hands us a perfect formula for The Thinking Man's Energy Diet. As in the grandfather clock, entropy should be generated fast enough to make things interesting—but not so fast we can't enjoy the ride. "All things in moderation" might be another way to say it. All that remains before transforming these impressions into a more scientific statement is to consider entropy from yet another vantage point.

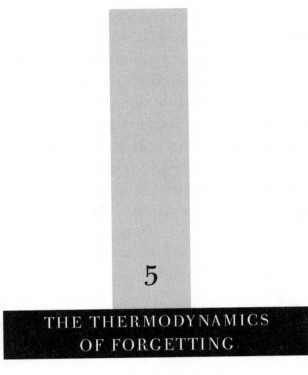

5

THE THERMODYNAMICS OF FORGETTING

The Energy in Information

In the last chapter, I referred to nature's punishment for a "fool-hardy" entropy decision without providing an example. From the context, you may suspect that acting on such a decision will generate a lot of heat, so starting a forest fire would likely seem one of these fool-hardy deeds. However, I was thinking about something subtler. The entropy producer I am thinking of has to do with *forgetting*.

The origins of the monumental discovery that forgetting produces heat (which has to rank among the greatest accomplishments of humankind) can be traced back to Ludwig Boltzmann (1844–1906), a corpulent and pugnacious professor of theoretical physics at the University of Vienna. Boltzmann wasn't concerned with the thermodynamics of forgetting; he was preoccupied with entropy and believed the path to its more complete understanding was by way of the atom.

In the mid-nineteenth century many of the most influential minds in natural philosophy thought of atoms as a useful paradigm but not as real things. Atoms were useful because many gaseous properties could be rationalized from thinking about atoms as tiny billiard balls. For example, if the length of the edges of a cube holding a fixed quantity of air is doubled, then the pressure in the cube drops by a factor of eight. This was explained by assuming atoms travel with a uniform velocity and bounce back and forth between the walls of the cube. The impacts cause pressure, which is proportional to the number of impacts per unit area per unit time. When the edge length is doubled, the surface over which these impacts occur increases by a factor of four, and the time between impacts—the atoms now have twice as far to go—increases by a factor of two. Hence, the number of impacts per unit area per unit time drops by a factor of eight. Voilà, volume and pressure are inversely proportional. This relationship holds regardless of the specifics of the atomic motion. Assume any reasonable motion for the atoms, and the result will be the same.

To some, it didn't seem to matter whether or not atoms existed; to others, it was a question of fundamental importance. The disagreement fueled a real brouhaha, with the whole philosophy of science at stake. The hot button was: Should science admit as real something that has not been detected through experimentation? Many felt that to do so would violate the time-honored criteria of a successful theory, which was to predict the results of an experiment. (If your theory called for atoms, you had better come up with an experiment to detect them.) Others believed that if theories built on the existence of atoms correctly predicted the things we could measure—such as the relationship between pressure and temperature—then there is sufficient reason to conclude that atoms are real.

The ideas of the great physicist James Maxwell (strong proponent of the atoms-are-real camp) caused Boltzmann to ponder the relationship between atoms and entropy. Maxwell visualized atoms mingling, colliding, rebounding, and fusing to form larger molecules. In the 1850s he wrote:

> I have been carried . . . into the sanctuary of minuteness and of power, where molecules obey the laws of their existence, clash together in fierce collisions, or grapple in yet more fierce embrace, building up in secret the forms of visible things.

Maxwell didn't have the time or patience to deal with little balls moving at the whim of a scientist's imagination. Atoms were real, and he wanted to know how fast they were going.

Like others before him, Maxwell assumed molecules were like miniature billiard balls. Unlike his predecessors, he used statistical methods to find their most probable speed distribution. He calculated the velocity distribution of gaseous molecules and soon realized that "live" atoms and molecules would possess a wide range of speeds. Though the First Law of Thermodynamics requires molecules in a closed container at constant temperature to move with an unchanging average kinetic energy, some molecules will hustle while, simultaneously, others piddle along. Maxwell wanted to know exactly how many molecules were going how fast. In other words, he wished to know what fraction of the molecules in a container at a specific temperature had speeds between 0 and 1 kph, between 1 and 2 kph, between 2 and 3 kph, etc., up to very high velocities.

He discovered that the average speed of molecules was related to temperature, but many moved much faster and much slower. To provide a point of reference, the average speed of a molecule of nitrogen or oxygen in the room where you now sit (except for those of you who are reading this in a sauna or in an unheated tent near the South Pole) is around 440 meters per second, or 1,000 mph. However, a very small number of molecules are hardly moving, and another small number have velocities in excess of 2,000 meters per second, or 5,000 mph.

With Maxwell's expanded billiard-ball model of gases, pressure resulted from atoms colliding with the walls of its container; heat was the total energy of motion of all the atoms in the container; and temperature, on the Kelvin scale, was proportional to the average kinetic energy of these atoms or molecules. (This is why things can't get colder than zero Kelvin. Once something's constituent molecules stop, kinetic energy vanishes. Negative kinetic energy is impossible.) As to the origins of entropy, there wasn't a clue.

Boltzmann wanted to know the "cause" of entropy. The dilution picture was useful, but it did not lend itself to predictions. For example, a gram of gas confined to a one-liter container at 300 K has less entropy than the same confined gas at 400 K, but the volume in which they are contained is the same. So, while this interpretation of entropy as diluted energy is functional, Boltzmann wanted to know *what* the energy was diluted in. He believed atoms were real, and that by understanding the motion of these undetectable physical entities, he would come to a better understanding of entropy.

Boltzmann's conviction invited constant conflict with some of the more influential physicists of the day, such as Ernst Mach, who held that atoms were nothing more than useful models of a much more complex—and perhaps incomprehensible—reality. Indeed, the war over atoms raged during most of Boltzmann's life. In 1905, Albert Einstein launched the final salvo with his explanation of Brownian motion: a phenomenon in which small dust or smoke particles, suspended in a still fluid, like water, move erratically. Sometimes the motion of these particles is described as a drunken man's walk, staggering one way and then another, making no real progress.

Einstein noted that if the fluid were made of atoms that obeyed the Maxwell speed distribution, at any given time more would collide with one side of the particle than with another, pushing it in that direction.

An instant later, there would be more collisions coming from another direction, and the particle would move accordingly. Using this model, Einstein was able to provide the first estimates of an atom's mass. Short of real atoms, there was no other way to explain Brownian motion.

Just a few months after Einstein's victory, on September 5, 1906, Boltzmann committed suicide while on holiday near Trieste, Italy. His daughter discovered him hanging from a hotel window, a lamp cord wound around his neck. There has been much speculation in the last century that the constant barrage against Boltzmann and his atomic theory led to his demise. While it may have contributed, it certainly was not the cause. Boltzmann had struggled with depression for decades.

During his life, Boltzmann did not let the opposition dissuade him from his search for the cause of entropy. As is often the case, Boltzmann's great contribution came by way of answering an elementary question. Whereas Maxwell had determined the most probable speed distribution for the molecules of a gas, Boltzmann thought about its odds. Let me explain:

Say there is a system of 100 tiny, stationary billiard balls. Because none of the balls are moving, the total kinetic energy of the system is zero. To this system, a unit of energy is added by setting the balls in motion. This can be done many ways. For example, all of the energy can be given to a single ball, and because there are 100 balls, there are 100 ways to add one unit of energy to the system. The energy could also be inserted into the system through two balls, each possessing a half-unit of energy. There are 100 ways to pick the first of these two balls and 99 ways to pick the second, giving $100 \times 99 = 9,900$ ways to distribute energy in the system (actually it is $9,900/2 = 4,950$ ways, because the balls are indistinguishable). For three balls, each with a third of an energy unit, there are 161,700 ways to distribute the energy; for four balls there are 3,921,225 ways; for five, 75,287,520 ways, etc. You can see that the numbers are getting larger very quickly. If one unit of energy is evenly distributed among 50 balls, there will be a whopping 100,891,344,545,564,193,334,812,497,256 ~10^{29} ways to distribute the energy.

It may seem that the greatest number of ways to distribute one unit of energy to the system is by dividing it equally among all 100 balls. However, there are not billions, or millions, or even hundreds of ways to do this. There is only one. Each ball must have $1/100$ of the energy.

What may appear counterintuitive in this example becomes obvious when stated another way. I have forced the balls in our system to be in one of two groups, where the first group is characterized by uniform speed and the second is stationary. Imagine now tossing one hundred pennies in the air. Some would land heads up; others, tails. We could link the penny-tossing and moving-ball experiments by associating the number of heads with the number of moving balls, tails with stationary balls. There is a one-to-one correspondence between the outcome of my thrown pennies and the moving-balls experiment.

So, if I were to ask for a prediction on the outcome of the penny toss, you should identify the most likely outcome—the "best bet"—as fifty heads up, fifty tails. The same best bet applies to a box inside which 100 balls are stationary or moving with a fixed speed. If you bet 50 moving balls, you would win approximately 8 percent of the time. That's not bad. By contrast, betting on 40 moving balls (40 heads and 60 tails), you would win about 1 percent of the time. And you are less likely to win betting on 20 moving balls (20 heads and 80 tails) than you are to win betting on Powerball.

However, people do win Powerball, and a 1 percent chance does not mean we will never see 40 heads after flipping 100 coins. Tiny balls and flipped coins only serve to represent the moving atoms and molecules of the real world. In it, a balloon holds more atoms than there are blades of grass on all the world's golf courses. As the numbers increase, the percent by which the outcome deviates from the best bet shrinks. When talking about the outrageous number of atoms or molecules that constitute what we can actually see, these deviations are essentially nonexistent.

Knowing the best bet is almost instinctual, yet articulating why it is so, is more complicated, which is why it took a genius to make the leap. To get where Boltzmann was, let's build one more mental model and imagine playing marbles in an orbiting space shuttle. We would bunch all the marbles together in the middle, and then shoot a single marble at the mix. In short order, marbles would be flying around the shuttle, colliding and rebounding. If we filmed the scene, after a few minutes we would not be able to tell if the movie was being shown in the forward or reverse direction. Because no single frame stands out as unusual, each frame is equally likely. However, as the movie continues to run, suddenly all of the marbles come together in a bunch and spit out a single marble. At that point, we would know the movie was running backward, because this arrangement of marbles is very unlikely. Ludwig Boltzmann knew that what we observe are the things that can be realized in the greatest number of ways—moving marbles that look the same going forward as reversed. So if we start a system out in a contrived manner—marbles all bunched together and hit with a single marble—the system will evolve to a state that can be realized in more ways, stopping only after finding the state that can be realized in the greatest number of ways. Boltzmann's realization: Energy is being diluted over the number of ways a system can be realized, and entropy is related to that number.

Remember $\Delta S_{univ} > 0$, the mathematical statement of the Second Law, which means that for something to happen in a reasonable amount of time there must be an increase in entropy. We now see that this is tantamount to saying that for something to happen, the final distribution of energy in the universe must be realized in more ways than its initial distribution.

Through brainwork, Boltzmann was able to deduce that entropy is proportional to the logarithm of the number of possible ways a system could be formed.[1] In symbols, $S = k \ln(W)$, where W is the number of ways to realize a system, **ln** denotes the logarithm of this

number, and k is a *proportionality constant,* now known as the Boltzmann Constant. This most famous of equations is all that appears atop
Boltzmann's headstone.

The depressed yet determined professor of theoretical physics
transformed entropy from something that had to do with heat into something connected with probabilities. Improbabilities can be realized in
only a few ways and have low entropy—a royal flush—while common
events are common simply because they can occur in many ways and
hence have high entropy—all the worthless poker hands. As straightforward as this concept is, scientists found a way to make the whole
matter seem much more confusing. Instead of counting the number of
ways a system can be realized, we talk about its "order." Things that
are ordered are said to have low entropy. The confusion arises because
order is seldom defined, leaving the nonscientist with only a subjective
sense of orderliness as a way of assessing relative entropy. Invariably,
the example of a bedroom or office is used to illustrate the Second Law.
The argument goes like this: Over time, an ordered (neat and clean)
bedroom tends toward disorder (messy and dirty), fostering the idea
that a particularly clean bedroom possesses low entropy and a particularly sloppy one has high entropy.

The problem here is that a specific clean or messy bedroom can
be realized in only one way—the way it is. Entropy only has meaning
within the context of the number of ways a particular macroscopic
state can be formed. In the example above, only two macrostates are
possible, ordered or disordered. (Though the distinction between the
two may be subjective, no matter what your standards—within reason—
there will always be many more ways to arrange the contents of a bedroom to make it messy as opposed to tidy.) *Order* is just another word
for *entropy.* It refers to placing things in sequence—a unique, low proba-
bility arrangement among all those possible.

It would be almost fifty years before another of entropy's many
hiding places was exposed. The game of hide-and-seek was playing

out in the burgeoning field of telecommunications. As telephones and telegraphs proliferated, a problem arose. Engineers needed to design communication networks, but they didn't know how to compute the channel capacity of these networks. Too small a channel resulted in an error-riddled, or "noisy," message. Then, in 1948, a Bell Labs scientist named Claude Shannon published a remarkable paper that spawned the science of *information theory*.

Shannon thought of information as "what you don't know." If you already know something, there is no need to communicate it. From his point of view, information was uncertainty. Networks had to transmit messages without regard to their content. In other words, they must be designed to handle all the messages that could possibly exist.

Shannon formulated an equation that related message uncertainty to the message's length. He didn't want to call this uncertainty "information," because he considered the term to be overused and misunderstood. As the story goes, the eminent scientist and mathematician John von Neumann noted that Shannon's equation was identical to Boltzmann's entropy equation, saying, "You should call it entropy, for two reasons. In the first place, your uncertainty function has been used in statistical mechanics under that name, so it already has a name. In the second place, and more important, no one knows what entropy really is, so in a debate you will always have the advantage." Thus, entropy and information became synonymous. If entropy measures the uncertainty of information, then knowledge is low entropy information. In other words, truth possesses low entropy—there is only one way (or at best, a few ways) to be right and many ways to be wrong.

This is much more than a game of semantics. Recall that heat generation is required to accompany entropy production. So if information entropy is the same as the entropy of Carnot and Clausius, then in going from a state of knowledge to uncertainty, heat should be produced. Guess what? This is exactly what happens. If I take a known message, say, one that is encoded in a computer, and erase it, some heat will be generated. No matter how perfect the erasing mechanism, the very act of erasing generates heat.* Dividing this heat by the temperature of the surroundings gives the thermodynamic entropy, which is identical to Shannon's information entropy.

If erasing information produces heat, then by the First Law, information must be a form of energy. Further, since energy and mass are the same thing, information must have mass. This is mind-boggling. A blank computer disk weighs less than one on which data is stored!

What is true of computer memory is also true of our memories; the act of forgetting produces a little heat and generates a little entropy. To understand the importance of this observation, let's take a closer look at the grandfather clock model of our universe.

Begin with a beam of light striking a barren earth. There it is absorbed and converted to heat. Next, instead of bare rocks and soil, imagine a wheat field. This time the light falls upon a leaf, and some of its energy is stored in the wheat starch. What remains becomes heat and is lost to the air, just as the Second Law demands. The grain of wheat, in which the sunlight's energy resides, ripens and falls to the earth. There, a mouse consumes it. Once more the sun's energy is transformed; this time the wheat's starch is used to synthesize muscle protein. What isn't used for this purpose becomes heat.

In the way of all things, a circling hawk swoops down on the mouse and the last little bit of energy from that sunbeam makes yet another transformation. The chemical energy from the mouse's protein is used to recharge one of the hawk's mitochondria; the remainder is lost to the Second Law. When next the hawk hunts, this enduring portion of the sun's energy will complete its journey to heat as the

---

*This insight came from Rolf Landauer in 1961. Sometimes we fail to appreciate that there is still much to understand about the "simple" laws of nature that were discovered centuries ago.

mitochondria supply energy to the raptor's mighty wings, which in turn heat the air.

The inevitable journey to heat took only an instant in the absence of life. In its presence, the process slowed, taking hours, days, perhaps even weeks before the energy in a single ray of light was converted entirely to heat. This is what life does; it slows the march of the Second Law, and I argue that those things that slow entropy production most effectively are rewarded. They survive. This is a key concept to all that follows, so let's take a moment here and explore what I think it all means.

What exactly is it that survived in the sun-grain-mouse-hawk vignette? Sure, the hawk came out on top this time. However, there are more mice, more hawks, more grain, and more sunlight, allowing many variations on this theme. Sometimes the mouse will escape and the hawk will die from starvation. Sometimes the grain will take root, avoiding the mouse altogether. Whatever the outcome, life goes on. And its survival is made possible by passing information—encoded in DNA—from one generation of grain to the next, from one generation of mice to the next, and from one generation of hawks to the next. Life has survived because information has survived.

Think of life not as individual organisms but as a communication network that transmits and stores messages. In other words, think of life as a DNA storage and transmission system. The Second Law tells us that at any given time, a specific amount of entropy is contained in the messages of this communication network. Because the transmission process is spontaneous (sex happens) and irreversible (sex doesn't *un*happen) the entropy in the network must increase. In other words, the number of different messages in our living network will grow larger. Starting with one given DNA sequence, it will evolve into many, and these will evolve into many, many more.

Some antievolutionists hold the position that Darwinian theory is a violation of the Second Law of Thermodynamics. On the contrary, in the living system of our experience, which propagates by the transmission of information, evolution is a *consequence* of the Second Law.

While the laws of thermodynamics mandate that there be a growing number of different DNA messages in the living world, the laws are mum on the content of these messages. However, the principle of natural selection—survival of the fittest—screams out on this topic. Remember, we have rephrased this principle as "survival of the thermodynamically most efficient"; that is, of the growing number of DNA sequences only those that make the living world more efficient—slow the rate of entropy production—get passed to the next generation. Natural selection acts as a filter in our communication network that blocks the transmission of specific messages. Together the Second Law and the principle of natural selection act to make the biosphere into a gigantic machine that remembers and learns ways to make each of its parts more energy efficient.

We are unique among living things in that we imagine—which is a spontaneous process. Hence the Second Law guarantees that the bounds of our imaginations will expand. We also discover and transmit knowledge. So voluminous is our knowledge base that no brain can hold it all. We have consequently developed a complex system to remember it: It is stored as hieroglyphs, scrolls, books, records, films, CDs, DVDs, magnetic memory, flash memory, and RAM. In our childhood, adolescent, and early adult lives we are largely dedicated to fulfilling a cultural duty to remember a portion of this knowledge. We call it schooling. Once we have moved this knowledge to our brains, we are in a position to transmit it to others, but not necessarily in its original form. First we may do work on it, brainwork. We may test it for consistency, accuracy, and its ability to contribute to an efficient *knowledge acquisition* system. Only if it aces these tests is it passed on. If it fails, it may be amended for accuracy or consistency, or it may be branded untrue and discounted. Thus we have created a filter system to check the veracity and usefulness of the things we imagine.

Our system of knowledge acquisition and transmission has much in common with the passage of DNA through generations. So just as life is a system that becomes more energy efficient—that is, gets more work and less entropy out of each of its parts—human beings are a system in which individual "parts" evolve toward greater brainwork efficiency.

The parts I refer to are constituted whenever people come together for a common purpose. Individuals, families, businesses, cities, nations are all parts of the human machine. Those parts that are more efficient at generating brainwork than their competitors are the parts that will survive.

Nature has spent billions of years converting the energy of the Big Bang into work and heat. The more ways it can find to do work efficiently, particularly work that results in long-lived products, the longer the universe can stretch out its life. First the cosmos made atoms, then molecules, then life, and from life came a special creature that could do work by accumulating knowledge. Of course, that creature is us. And the knowledge we accumulate is the last step in energy's descent from the Big Bang. When it is forgotten, the passage to heat is complete.

We have a responsibility to nature to use a portion of the energy we consume to remember, do brainwork, and learn. And the most important lessons to learn are those that will allow us to be more efficient with the universe's energy. In my opinion, that is why we are here, and to forget this is a foolhardy entropy decision.

Our examination of the laws of thermodynamics has taught us that simply reducing energy usage will not be sufficient for our diet. We must simultaneously increase our output of brainwork. The adage from weight-loss diets, "Eat less, exercise more," has a counterpart in energy diets: Less energy in, more brainwork out. Now we need to examine where the "energy in" comes from.

Energy Metabolism

6

THE ENERGY RIDE

The Big Picture of Energy Metabolism

As a child, I had two energy obsessions. The first was chemical energy, which fascinated me. I have a dim memory of sitting alone on the floor in our kitchen, holding a book of matches. I was trying to strike one, but in the process burned my fingers. I have a much more vivid memory of having my hand slapped when I was found playing with an expropriated cigarette lighter. From this experience I concluded that fire, and by extension, explosives, were an adult activity that I would "grow into" like shaving. However, I was an impatient child and, preferring not to wait, took it upon myself to self-educate.

With my mother's full consent, I constructed my first "bomb" at age six, having mastered my match-lighting skills a few years earlier. To be fair, when I asked if it was okay to build the bomb, Mother appropriately requested further details (which I was happy to provide—to a point), and then she said, "That's fine." I told her that I was making my bomb from a can, steel wool, some wires, and a battery; what I didn't mention was that it was a gallon can, containing at least a cup of something that smelled very much like gasoline.

The results were both spectacular and unexpected. I positioned my would-be explosive in the nearly empty brick garage. There I would detonate it from the outside, where I would be protected from the inevitable blast. With a crude but effective igniter, made by including steel wool in a circuit powered by a nine-volt battery, my plan was set in motion.

But there was no explosion. Just an elementary whoosh followed by a metallic twang that repeated its song every few seconds. As I peeked into the garage, I saw the can undergoing a series of expansions and contractions; a brief geyser of flame, shooting about a meter into the air, accompanied each expansion. A short quiescence followed, after which the can contracted, producing the now familiar discordant sound. Then the sequence began again with another pyrotechnic belch. Though I was disappointed there had been no explosion, this sequence of flame and animated sounds was way cool.

Just as I was beginning to wonder how I'd stop this fireworks display, my mother came running from the house, evidently in response to my big sister's screams. In what I considered a serious lapse of judgment, she kicked the combusting can onto its side. Propelled by horizontal flames, the can began to spin like a top. Mother was now forced to play a potentially hazardous round of hopscotch with this rocket-propelled, flame-spewing gas can. Luckily, she only had to make a couple of nimble leaps before the contents were exhausted and the show concluded. No serious injuries were incurred, and I lithely escaped punishment by reminding Mother that when I had asked permission to build the bomb in the first place, she had answered . . . "Fine."

I learned two things from this experience: to be more covert with future bomb-building activities, and stuff that burns doesn't necessarily explode—contrary to what I had surmised from watching *Peter Gunn* and *The Untouchables.* There gas tanks exploded whenever a car left the road, particularly at the top of ragged, rocky cliffs. Had I been older, wiser, and less fearless, I would have learned something more urgent that day—but for now, that lesson would have to wait.

My second obsession was roller coasters. Like matches and explosives, this preoccupation was probably exacerbated by a perception that it was yet another activity I had to grow into. On my first trip to an amusement park at age six I was denied a ride because I was too short (according to a four-foot green-shod gnome painted on plywood holding a sign that said, "If you are not taller than this, you cannot ride"). Several years passed, but when the time finally came, I climbed into the very last car of the train. I'd heard that was where you got the best ride.

Having never ridden a roller coaster before, I couldn't say whether the last car was the best place or not. But I could attest that it provided an exhilarating adrenaline rush. From the coaster's initial descent *(How had I gotten myself into this situation?)* to the instant when the coaster stopped and I rummaged around for another ticket, I was hooked. There was no line for the coaster that day, so I made six consecutive circuits, in the last car every time. As I handed over my ticket for the seventh pass, the attendant suggested I try the front. "The whiplash is better," he said.

Better? How could anything be better? Yet without hesitation I moved to the very first car. Certainly, the stomach-churning sensation from the uninterrupted view of the first drop was intensified, but the physical sense of falling was unchanged. And, as I did on the six previous rides, I began to laugh. My mouth thus wide open, the coaster reached the lowest point of its descent and began climbing the next hill. The radial acceleration, greatest for the leading car, produced a downward force that conspired with my laughter, slamming my jaw shut. Unfortunately, my tongue got in the way of my teeth, and off came a rather respectable chunk of the former. Blood ran down my chin and onto my shirt, and this time when the coaster came to a stop, I had had enough. As I headed toward the exit, the attendant handed me a free pass, and believe it or not, thirty minutes and three snow cones later, I used it.

Explosives and roller coasters, though quite different, are controlled by the same principles of nature, including the Second Law of Thermodynamics. The Second Law mandates that the energy of these devices be diluted into the surroundings as heat. A proper explosive transforms potential energy into heat just as roller coasters do. The energy in my gas-can bomb was transferred as heat to the air in and around my parents' garage, and all of the energy imparted to a coaster as it climbs the first hill is converted to heat—in the wheels, track, and air—by the time it comes to a stop. Despite these similarities, it would be unwise to substitute explosives for a roller coaster, or vice versa. Even if the plywood gnome of my youth came to life to dance taunting circles about me, I would still never be tempted to hop on a ride powered by an exploding stick of dynamite. And it would be silly to use a roller-coaster's power to excavate a mineshaft. What makes one device unsuitable as a substitute for the other is the time required to convert potential energy to heat. In an explosive, this conversion is completed in an instant; on a coaster, minutes.

The art of roller-coaster design is all about slowing entropy production. Instead of converting potential energy to heat, it is transformed to kinetic energy (at the bottom of a hill) and then back to potential (at the top of a hill). This sequence is repeated again and again. However, like the fee charged when foreign currency is exchanged, nature charges a heat tax with each "transaction," taking the entire initial investment in the long run. With the aid of hard, well-lubricated ball bearings, steel wheels, and unyielding tracks secured to a rigid platform, the tax rate can be reduced. But still, the inevitability of the Second Law ultimately brings the roller coaster—as well as every other process—to a halt. Indeed, if our current best bet goes unchallenged, the universe itself will stop when the initial flash of energy from the Big Bang is converted to a uniformly cold expanse of darkness. But how soon that happens depends on whether the universe unfolds like a stick of dynamite or an amusement park ride. Since it is now 14 billion years old, it obviously didn't explode into oblivion—despite the image the words *Big Bang* conjure. Instead, it must be more like a roller-coaster ride. And while it has its ups and downs, like a coaster, this ride has some significant differences.

For starters, the universe does not run on a closed track, with each car following the same path. Instead, the cars of this "energy ride" can follow multiple and different routes to the bottom, much like at excellent ski areas where there are dozens of trails from summit to base. Secondly, the thrill of the ride is in moving from one form of energy to another, not from place to place. On this ride, a rapid descent involves a *loss* of total energy, which, by the First Law, mandates that something else *gain* that energy. One way to maintain this balance is by coupling and decoupling the cars to form longer or shorter trains. For example, a train plummeting from ten to five joules would lose five joules. These have to go somewhere. One possibility is that cars of the rapidly dropping ride arrive at the five-joule level as two shortened trains. The total energy of these two sets of cars is ten joules, and, in concert with the First Law, energy has neither been created nor destroyed.

Why does this matter, from our selfish anthropomorphic perspective? In the process of decoupling, shorter trains are often left behind, stranded in shallow valleys far from the ride's end. On its way to becoming heat, the universe is leaving behind little bits of itself as potential energy. With just a nudge, these bits can pick up where they left off along the road to heat. But now, 14 billion years in the making, *we* control the route followed—just like we can mastermind the falling weights powering a grandfather clock. By controlling the way the weights fall, we can get something out of the process—the time of day, the phase of the moon, melodic chimes.

The cars of the universe's energy ride travel in six dimensions, what physicists call *phase space*. Fortunately, we can develop a useful representation of this ride without sporting multiple pairs of 3-D glasses. Begin by picturing yourself in a hot-air balloon floating miles above a vast countryside of many mountains and valleys. At its center is an immense peak surrounded by a deep moatlike canyon. The topography is suggestive of a Mexican hat. So immense is this central mountain that, even from many kilometers away, you must tip your head back to see up the steep angle to its mist-shrouded summit. Moving away from this exaggerated sombrero is a jagged landscape that grows more undulating as it transforms into a hilly expanse, finally fading into a gentle descending wasteland that stretches to the horizon. Spiraling about the central peak, called Mount Genesis, is a mountain range consisting of about a hundred peaks separated by valleys of varying depths. As you travel along the first quarter of this range, each successive valley is lower than the one before. From that point on, however, the valleys ascend slightly, reaching a high point somewhere near the end of the range. The runoff from rain or snow falling on this landscape form many small streams, winding from valley to valley before merging into rivers and flowing into the wastelands.

There is no moisture and no vegetation here; for that matter, there are no rocks, alpine forests, trailheads, or ski lifts to provide contrast to the white background of the peaks and expanse beyond. Wishing for a better look at the stuff from which this energy ride is constructed, you maneuver your hot-air balloon above a large peak and begin to descend. Just before touching down, white blocks (from which the entire landscape is built) come into focus. What appeared to be smoothly inclined slopes are actually Lego-like blocks exposing sharp horizontal steps and vertical faces. The cars of the universe's energy ride do not so much roll as hop from step to step; each step's elevation supplies the total energy for the car or cars that rest there. To make matters more complex, only certain combinations and kinds of cars can occupy each step.

Take a giant mental leap and don't think of a step, or group of steps, as a place, but rather as a potential thing. Cars trapped on the steps of the deep moat at the bottom of Mount Genesis will take the form of hydrogen atoms. Cars in successive valleys will appear as atoms of each of the approximately hundred known elements, from which this imaginary string of mountain takes its name, Element Range. Those in the second valley will appear to be helium, the second atom in the periodic table; those in the third valley will take the form of lithium, etc.

But there is more to our universe than atoms. There are steps for everything that has been, can, or will be. For example, some steps are currently occupied by the energy that is you (remember, mass is just another form of potential energy). A more numerous set of steps forms the Sun. Similar groups of steps are now unoccupied, but in the future the cars of the energy ride may come to rest upon these and form a new star. There are steps for galaxies and would-be galaxies, for the atoms that form the pages of this book, and for pages yet to be and never to be written. Because cars at rest give existence to things with a fixed total energy, there are different sets of steps for a slow-breaking curveball and a fastball; a car fixed to a step of the energy ride does not necessarily form a motionless thing in the real world. Clearly, the number of steps making up the energy ride is an obscenely large figure, and at any one time cars occupy a vanishing fraction of these. However, recall that the First Law requires that the sum of the energies (elevations) of all these occupied steps must be constant. A group of cars can move to a lower step only if another group moves up, or it can move by uncoupling and occupying several lower steps. In either case, the sums of the elevations of occupied steps must remain unchanged by the move.

The Second Law of Thermodynamics also makes an appearance on the energy ride. At the start, all of the cars are stacked neatly atop Mount Genesis. The ride can only proceed if these cars uncouple and descend this first peak to become the hydrogen atoms filling "Hydrogen Ravine." It is not gravity propelling the energy ride's cars down the central peak but rather the Second Law. There is only one way for the cars to rest on the central peak but many ways to come to rest at the bottom. Moving down and out on the energyscape corresponds to an increase in entropy, as smaller "chunks" of energy are diluted over a greater number of steps. Entropy is at its maximum when single cars are distributed randomly among the many steps of the wastelands, stretching far beyond the horizon. This is where energy would go to die—if it could! But since energy can't be destroyed, it spreads out into nothingness, becoming increasingly thinner and more dilute as it expands into the wasteland flats.

At this very instant, you are a passenger on the universe's energy ride. You are now making your way over the bumps just beyond Element Range, and though you have no memory of navigating the twists, turns, and drops that came before, rest assured the energy that is you has been a traveler since the very beginning of time. And that energy will continue its ride for billions of years to come. Still, we are not allowed to experience the entirety of the ride in our human form, so let's do the next best thing and imagine what it would be like. A task that will be a little easier if the ride is made from high-tech Legos.

In your mind's eye, dim to blackness the lights making the energyscape visible. Then, imagine that when a step is occupied by a car, an internal LED (light-emitting diode) switches on, causing it to glow like the keys of a cell phone. The LED is extinguished when the car moves on. As a result, the progress of the energy ride is traced in blinking lights. The universe becomes an incomparable light show, its history and evolution revealed in the flashing of countless LEDs.

Fourteen billion years ago, the glowing LED atop Mount Genesis provided the first indication that the energy ride was to make its inaugural, perhaps only, run. The cars were stacked neatly on the highest step of this peak. Under the influence of the Second Law, this improbable array collapsed, time began, and the stuff of the universe formed as some of the cars uncoupled and fell from the mist-shrouded heights, tumbling in all directions, past cliffs and outcroppings. Some sections landed on these ledges, taking the form of peculiar, short-lived particles, which soon came apart and fell farther down the mountain. The exact properties of these particles are a matter of conjecture and the subject of high-energy physics. Regardless of the specifics, from our hot-air balloon, the first few seconds of creation appear as if someone had scattered a huge bonfire down the slopes of a very steep mountain. The burning debris passed down its sides as an encircling doughnut of light, fragmenting into smaller pieces all the while, until its LED embers settled like a thin blanket of gauze on the walls near the bottom of Hydrogen Ravine, leaving the peak in total darkness.

After a spectacular first few seconds, the next million years or so were relatively dull. The light emitted by Hydrogen Ravine continued, looking more like effervescent dust than the primary constituents of a dynamic exploding universe; the only change was the slow but relentless concentration of the ride's cars near the bottom of the ravine. At the same time, a very small percentage of the Legos forming the wastelands—what we will now call Photon Flats—began to turn on and off. With each passing eon, however, the interval during which these lights were off grew shorter. And then, 300,000 years after the energy ride began, the cold light coming from these diodes stopped flickering and remained on.

When it appeared that the show was about to settle into repetitive monotony, with some lights steadily aglow on Photon Flats and those at the bottom of Hydrogen Ravine permanently switched on, something began to change. The glow, uniformly distributed around the bottom of the ravine, began to coalesce. While the total number of "on" LEDs remained the same, they began to meld into huge glowing clouds and within these smaller, highly concentrated regions of illuminated LEDs formed. These were the first galaxies and stars. The greatest number and concentration of "on" Legos corresponded to the largest stars, which almost immediately began to elongate in the direction of the ravine's outside wall, that is, toward the sombrero's brim. Upon reaching the highest point, the LEDs in Hydrogen Ravine began to switch off as some of those in adjacent Helium Valley switched on. The energy lost as cars fell into Helium Valley appeared as spectacular light cascades sweeping across Photon Flats. These cascades began high on the gently sloping terrain and then moved in an expanding arc of illumination toward its limitless edge.

Meanwhile, back at Hydrogen Ravine, smaller stars were now pushing their way up and over the brim of the sombrero into the valley below. From here, through some unseen conduit, the cars of the energy ride began to bleed into valleys and dales even further into Element Range. LEDs in Carbon Dale—valley 6—switched on as some of those back in Helium Valley went dark. Then light began to pour forth from Oxygen, Neon, and Silicon Valleys. Soon, from a cosmic perspective, LEDs brightened all of the lowlands between Hydrogen Ravine and the lowest valley along the range, valley number 26—Iron Basin. The first and lightest twenty-six elements of the periodic table had been formed. As the circle of light surrounding Mount Genesis fell to lower energies, compensating light cascades on Photon Flats grew more numerous. Yet despite their large numbers, these expanding arcs of light were utterly lost in the sheer immensity of the flats.

Let's take a moment here to remind ourselves that what we are seeing from the soaring balloon looks quite different in ordinary space. A concentrated region of "on" LEDs in Iron Basin does not correspond to a star of iron; rather, we are looking at the iron in a star made of many elements. In each of the other valleys, a circle of light shows the amount of that corresponding element in the star. In essence, a "thing" in ordinary space is spread out by the energy ride. This is important because as we look at the largest spot of light in Iron Basin, something unexpected happens. The spot begins to come apart as billions and billions of LEDs, separating from one another as darkness fills the region between. At precisely the same time, the other spots, which are the hydrogen, helium, and other elements in this star, also begin to spread outward in their respective valleys. The star is exploding, distributing the elements from which it is made across the galaxy.

The careful observer would notice that accompanying this explosion, lights in the valleys beyond Iron Basin—toward Mount Plutonium—switched on. Some of these remained on for an instant and some for millions of years before beginning a return trip down to Iron Basin. And, as always, the cars descending through the energy ride spawn light cascades that briefly illuminate the surrounding terrain before becoming too diffuse to be of consequence against the voracious darkness of Photon Flats.

Through this and other stellar explosions, the dim glow of galactic clouds is no longer confined to the base of Mount Genesis. It now extends across Element Range. And from this glow, stars continue to condense, though now condensation is also occurring across the gently undulating downs beyond Element Range. In the small declivities in this part of the energyscape, the cars of the energy ride become molecules—from which everything we can touch and feel is made. As lights switch on across Molecular Downs, silica (sand), water, carbon dioxide, and methane form. These "common" molecules further condense to form dust, planetesimals, and finally full-fledged planets. Like an aerial nighttime view of the Las Vegas Strip, huge sections of Molecular Downs are suddenly made visible by innumerable gleaming lights. Some of these blink on for but a second and then go dark, as lights in the deepest recesses of the downs appear to turn on permanently. And though this is far beyond our ability to verify, it appears that the changing lights are consistent with the First Law of

Thermodynamics—as we watch, on average, the lights on the downs move to ever lower elevations as their numbers increase.

Then, from a previously dark corner comes the unmistakable green iridescence of a single LED. What's odd is that it sits high in the Molecular Downs on a mesalike prominence surrounded by lowlands. Then another light appears, followed by another and another. As we watch the growing illumination, we note that it is being fed by cars of the energy ride that had come to rest in some of the deepest depressions of the downs, those corresponding to water and carbon dioxide. Through some unknown mechanism, these cars are gaining energy to scale the mesa. Though it is tempting to suggest that this region is not subject to the First Law, careful observation comes to the rescue as we note that each time a new light appears atop the mesa, an LED on Photon Flats is extinguished. Some of the molecules on this mesa use the energy of light to synthesize other "high-energy" molecules—the building blocks of life.

In real time, we have been watching the cosmos unfold for more than ten billion years. In that time, the cars of the energy ride have moved steadily away from Mount Genesis. Though the process is a slow one, a noticeable dimming of light is seen coming from Hydrogen Ravine. Clearly, it is only a matter of time before it is left as dark as Mount Genesis.

We have been so mesmerized by the light show that we have ignored the darkness creeping ever outward. First it took Mount Genesis, and it will eventually overtake Hydrogen Ravine and then Helium Valley. Ultimately, it will sweep its way across Molecular Downs. Life on the unusual mesa will pose no more resistance to the advancing darkness than does a sand grain to a tsunami. The boundary between light and dark, like the rings on a tree, charts the age of our universe and the inevitability of the Second Law.

But let's not look too far into the future. The universe is still young. Ninety percent of the original material found in Hydrogen

Ravine is still there, and the darkness that will inevitably sweep us aside is many billions of years in the future. Right now we have more immediate and manageable issues to confront—not the least of which is making the best use of what comes from the energy ride. To do that let's look a little closer at those shallow valleys where the energy ride's cars are stranded.

7

THE AGE OF ATOMS

We have glimpsed a cold, dark, uniform future. Lucky for us, the valleys and depressions of the energy ride have postponed this inevitability by trapping bits of energy before they reach the vast wasteland of Photon Flats. As long as there is energy in these low spots, and as long as we possess the necessary knowledge to nudge it along its way on a course we pick, there will be energy to fuel our lives and our brainwork.

Consider the first, and perhaps largest, of these energy traps, Hydrogen Ravine. A few seconds after the Big Bang, most of the cosmos's original energy resided here as a very hot soup of electrons, protons, and neutrons, as well as more exotic ingredients such as pions and neutrinos. The common ingredients are the most interesting to us, since these are the building blocks of atoms.

Currently, processes in and among approximately one hundred different kinds of atoms satisfy the majority of humankind's energy needs. In turn, atoms are built from just three basic particles: electrons, protons, and neutrons. The protons and neutrons pack together to form a very small but dense structure called the atomic nucleus. Within the nucleus, the number of protons determine which of the atom types, or elements of the periodic table, is formed. Hydrogen, for example, has an atomic number of 1, which means it has a single proton. Carbon—atomic number 6—has six protons, while plutonium atoms have 94.

Protons have a positive electric charge and so repel one another. It would take a tremendous amount of energy to force 94 protons together to form a plutonium atom if not for neutrons. While neutrons have no electric charge, they are attracted to protons by the "strong force," which is many times more powerful than the electrostatic force pushing protons apart. However, this force only makes itself felt when protons and neutrons are very close together. Neutrons are the mortar for protons. Those 94 protons of a plutonium atom need about 150 neutrons to hold them in place to form an atomic nucleus. Two atoms of the same atomic number but with a different number of neutrons are called isotopes. Particularly important to us are the isotopes of hydrogen called deuterium and tritium, which, in addition to a single proton, are characterized by one and two neutrons, respectively.

An atom's nucleus is located at the center of what is essentially a sphere. And though it accounts for more than 99.9 percent of the atom's mass, it occupies an infinitesimal part of its volume. In a mental image, assume that an atom is the size of a domed sports arena. Its nucleus would be smaller than the head of a pin. The intervening space, between the central pinhead and the stadium's walls, would be filled with the exact same number of electrons as there are protons in the nucleus. An electron's mass is only $\frac{1}{2000}$ of a proton's or neutron's, and it possesses a charge of the same magnitude but is the opposite sign of a proton—it is negatively charged. Indeed, the force confining an electron to the region around the nucleus is the electrostatic attraction among these particles of opposite charge. With an equal number of electrons and protons, an atom is electrically neutral.

It is convenient to think of an atom's nucleus as stationary with electrons moving about it in the quantum mechanical equivalent of an orbit called an orbital. Just like a satellite orbiting the earth, the exact trajectory it follows depends on its velocity and hence its energy. The faster a satellite moves, the greater its distance from Earth. In the same manner, an electron's orbital varies with its energy. Beyond energy considerations, however, the analogy between electrons and satellites erodes: The motions of celestial objects are determined by classical mechanics, but the laws of quantum mechanics govern moving electrons.

Because quantum mechanics seem to contradict everyday experience—for instance, something being in two places at once—people use a number of useful mental images to visualize an electron's motion. The image I prefer involves picturing the electron as "spread out" to make a kind of fluid. An orbital constrains this fluid to fill some parts of the atom's volume more completely than others. Each electron fills a different, though possibly overlapping, region around the nucleus. Consequently the amount of electron fluid varies throughout an atom's volume.

Imagine taking a one-liter container to various places in an Astrodome-sized atom and filling it with the electron fluid found there. We could then weigh this fluid to determine what fraction of an electron it represents. At a point near the nucleus, where the electron fluid is the most concentrated, one liter might hold one-tenth of an electron. Near the walls, where the fluid is more diluted, it might contain a thousandth of an electron. This varying concentration of electron fluid is called the electron, or charge density, and it is found through careful measurement or calculated using quantum mechanics. Two atoms of the same type and having the same energy will be characterized by the same charge density.

As with all models, this one must be applied judiciously. If the picture were completely accurate, it would be possible to fill a scoop with some charge density and take it to another place. But this is not the case. Electrons are indivisible. To move charge density around, all of the electron fluid in one orbital must move to a different one, which alters its energy and simultaneously will cause the atom's total energy to change.

One way an atom's energy changes is through the absorption of a photon of electromagnetic radiation, that is, visible light. The energy of an incoming photon is transferred to an electron and makes it move faster, causing it to change orbitals. However, only a discrete set of orbitals is available to an electron. So, by the First Law of Thermodynamics, if this photon is to be absorbed its energy must exactly match the energy difference between the orbital the electron is in and the one it will move to. When these energies match, the photons are absorbed; when they do not, the atom is transparent to photons of this energy. This is one reason why things like jewels are colored. Emeralds are green because one of the atoms from which it is made, chromium, is transparent to green and blue light but absorbs red light. The electron orbitals around chromium have energy differences equivalent to the energy of red photons.

The procedure also works in reverse. An electron sometimes moves to an orbital of lower energy, where it moves more slowly. The First Law requires that energy to go somewhere, and often it goes into making a photon of a specific energy, e.g., color. This is what makes mercury and sodium vapor streetlights work.

In the 1960s, streetlights gave off a sickly blue-white light because each contained mercury vapor and little else. Inside the bulb were two electrodes separated by a small gap. At a sufficiently high voltage, electrons can be made to jump that gap, bumping into some mercury atoms in the process. Occasionally the impacts knock electrons into higher-energy orbitals, leaving their original orbital empty. When any of the higher-energy electrons fall into this empty orbital, light is given off. For mercury, the energies of these falling electrons correspond to the blue-white light. Modern sodium vapor lights give off a warmer, yellow-orange glow.

The science behind streetlights makes sense to me now, but when I was a kid it mystified me. The first indication of any mystery at all occurred when my science teacher gave me a small handheld spectrometer. These are neither complicated nor expensive instruments, consisting of a longish cardboard tube with a tiny slit at one end and a small plastic or glass window at the other. The window is covered with an opaque coating and thousands of tiny parallel lines are etched into it, revealing the transparent substance below. A light source, viewed through the slit, is decomposed into a color spectrum, and the warm natural light of the sun becomes a rainbow. The cold light of a streetlight is broken into discrete threads of color, of which the four most intense are dark blue, aqua, green, and yellow.

I sat on our front porch in Denver for hours into the evening, straining to see the colors hidden in that ghostly glow of the mercury vapor lamp. They couldn't be seen with my naked eye, but then a quick look through the magic tube and there they were again, always those same four colors. Point the instrument at a neon light and, Wow! There were hundreds of light ribbons, shades of red, green, yellow, blue—every color I could name but many more I could not.

I soon learned that every element emits a unique spectrum. Scientists use this fact to identify elements, a technique I applied in a limited way, picking out the gas in a neon light. (Neon lights aren't necessarily filled with neon; some contain xenon or krypton. The spectrum of neon contains more red, xenon more green, and krypton more blue lines.)

Most bewildering was the spectrum of the Sun. Though at first it appeared to be a continuum of color—red fading to yellow, yellow to green, and green to blue—upon closer inspection I saw dark threads in this rainbow. Some colors were missing, which, it turned out, was the consequence of gases in the sun's corona absorbing light, causing electrons to move to higher-energy orbitals. When these electrons fell back into the empty states below, the emitted photons took off in all directions, and few, if any, ended up in my spectrometer. These dark bands can be used to identify the absorbing elements. In the case of the solar spectrum, it's mostly hydrogen and helium.

The same principle applies to other stars. With the aid of a telescope, starlight can be concentrated, passed through a spectrometer, and the elements in its corona can be identified. I remember excitedly trying to use my spectrometer for this purpose, but starlight was just too weak. However, with more powerful instruments we can literally look back in time to when the first stars formed and then, using a piece of etched glass, see the elements as they were born.

Recall that seconds after its start, the cars of the energy ride in chapter 6 settled into Hydrogen Ravine—a soup of electrons, protons, and neutrons. Atoms had not formed yet because these building blocks were very hot—many billions of degrees hot—so they all moved about independently. Electrons were not confined to the region around a single proton. Instead, they zipped around as "free particles," first whizzing past one proton and then past another at speeds approaching that of light. Some energy had to be shed from this soup before nuclei, and then atoms, could form.

Under ordinary circumstances, there would be no mechanism through which to cool the universe. While fast-moving atomic building blocks occasionally bumped into one another and slowed down a bit, energy conservation required the kinetic energy lost in these collisions to be exactly balanced by the emission of electromagnetic radiation. This radiation caused the soup to glow, like the burner on an electric stove. However, the light was not a dull red or even white; rather, these collisions produced x-rays, the component of the electromagnetic spectrum that has many times more energy than the part we see as visible light. In turn, these x-rays then smacked into protons, neutrons, and particularly electrons, which reabsorbed them to move faster and regain the lost kinetic energy. It is here where the universe would have ended, with x-rays being emitted and reabsorbed in an endless cycle, the stuff of the universe rolling like marbles in the brim of a sombrero. The expanding cosmos provided nature with a way out of this perpetual monotony.

It's tempting to think of this expansion in terms of an exploding bomb with shrapnel shooting from a nucleus, but that's an inaccurate interpretation. Instead, think of a piece of graph paper where each grid mark is one millimeter square. Leave this paper for a few billion years and you will return to find that the grids are now two millimeters square. The paper has expanded proportionately; any picture drawn on the graph paper would look the same, only bigger.

Similarly, as our universe expanded, the average distance between its parts—electrons, protons, and neutrons—grew larger. Work was being done against the mutual gravitational field of these particles and radiation. In compensation, thermal energy decreased and the universe cooled.

Just thirty-four minutes into this expansion, after the temperature had fallen to 300 million degrees, some of the protons and neutrons had shed sufficient energy so that the strong force could now hold them together. The result was a "plasma" of light atomic nuclei, principally deuterium, tritium, and helium (two protons and two neutrons), moving about in a sea of even faster free electrons. The cooling expansion continued for eons, and 300,000 years later the cosmic thermometer registered a moderate 3,000 K (about 5,500°F). At this temperature, electrons no longer retained the energy to escape the attraction of nuclei, so they joined them, forming a rarefied gas of hydrogen and helium atoms that filled a universe only one-thousandth its present size.

Atoms are far less likely to absorb photons than are free electrons, so when atoms formed, the universe became transparent. The glow—photons—from the 3,000 K gas filled all of space as it continued to expand and cool. Today, what is left of the glow can be "seen" with the aid of a television that receives a broadcast signal.[2]

Once again, the universe had come to a critical stage. By all rights, the atomic gas should have continued to expand and cool forever; that is, if its density were uniform. However, space was lumpy. We now know—though not why or how—there were just a few more hydrogen and helium atoms in some parts of the cosmos than in others.[3] And, where atomic density was the greatest, gravity pulled more forcefully at the surrounding material. Density variations amplified as newly formed atoms accelerated toward these gravitational centers. Around each, the density of hydrogen and helium grew; atoms collided more frequently, became hot, and began to glow red; and still, gravity tugged at this luminous gas, further compressing and heating the newly formed protostars.

While gravity tugged inward, the increasing temperature of these hot protostars produced an escalating outward pressure, which in some cases was sufficient to counter gravity altogether. However, for massive protostars the collapse continued. Gaseous temperatures approached 10 million K, well past the point where electrons were stripped from nuclei to form plasma. Now a reasonable fraction of the hydrogen nuclei were moving fast enough to get inside the repulsive electrostatic barrier that pushed these same positively charged particles apart. Once inside, they succumbed to the strong force, accelerating toward each other like kids playing at bumper cars, although, when

these nuclei collided, they did not bounce apart but instead fused together to form a heavier nucleus. Neutrons were attracted to protons to
make deuterium and tritium, which fused together to produce helium
nuclei. Helium nuclei are less massive than the combined mass of the
hydrogen isotopes from which they are formed, which is only possible
if some of the nuclear mass is converted to energy—and that's exactly
what happened. When the hydrogen became helium, a huge amount of
energy was liberated, some of which found its way to Photon Flats. But
a much greater part went to further heat the already white-hot gas. Now
more hydrogen nuclei had enough kinetic energy to move inside the repulsive barrier and fuse, making more heat, which went to drive more
nuclei inside the barrier, and so on. The conversion of hydrogen and
its isotopes to helium had become self-sustaining. Stars blazed to life.

While the first stars fed on hydrogen, deriving energy from its
conversion to helium, nature provided a multitude of energy-liberating
reactions from successive fusion reactions. Indeed, fusion to produce
elements as heavy as iron (26 protons) and nickel (28) releases energy.
However, extreme conditions are necessary to harvest this energy. Remember that the nuclei of hydrogen and its isotopes possess a single
proton and therefore have a positive charge of one. Helium nuclei,
with two protons, have a positive charge of two. Hence colliding helium nuclei repel each other more strongly than do those of hydrogen,
which means that the repulsive barrier blocking helium fusion is more
pronounced than the hydrogen fusion barrier. Simply stated, helium
nuclei must collide with greater relative velocity before getting close
enough for the strong force to dominate. This requires helium to be
much hotter than hydrogen before fusion becomes self-sustaining. In
most stars hydrogen fusion is spontaneous at 10 million, helium fusion
occurs at much higher temperatures—closer to 100 million K.

One of the more important fusion reactions of helium is the *triple-
alpha process,* where three helium nuclei, each with two protons and

two neutrons—also called alpha particles—fuse to make a carbon nucleus (six protons and six neutrons), which is the element upon which all life is based. The products of helium fusion then undergo subsequent fusions to make elements of progressively higher atomic numbers, such as oxygen, magnesium, silicon, sulfur, and argon. In each instance, higher temperatures are necessary to surmount the repulsive barrier. The synthesis of argon (18 protons and 18 neutrons) from sulfur (16 and 16) and helium (two and two) is significant at temperatures approaching a billion Kelvin. There is no temperature great enough to produce a self-sustaining fusion reaction to produce elements any heavier than iron and nickel. Beyond nickel, fusion consumes energy by manufacturing products more massive than the nuclei from which they are formed. This is a case where energy is converted into mass.

Just after the Big Bang, when everything was still in Hydrogen Ravine, and the temperature was well over a billion degrees, why didn't everything just fuse to iron? To fuse, two or more light nuclei must collide with sufficient energy to get inside the repulsive barrier. Temperature is a measure of the kinetic energy of individual particles but not of how often they collide. Collision frequency will depend on the particle density. Just after the Big Bang, though the temperature was very high, the universe's contents were distributed almost uniformly throughout its entire volume. This meant density was low and collisions were infrequent—which was how helium was formed. The chances of three helium nuclei colliding to make carbon were almost nonexistent. Think of it this way: You occasionally see a three-car collision, but usually where the density of cars is high, like at a Stop signal or an intersection. On an infrequently traveled country road, where the density of cars is low, you may never see a collision involving three cars. The same applies to nuclei. Before the universe could get serious about the manufacture of elements, it needed to concentrate atom-making ingredients into stars. And before stars could form, particles in

Hydrogen Ravine had to cool to just a few degrees above absolute zero—only then could gravity tweak the density of the universe's bits and pieces.

Let's return to the little detail that nickel and iron are the heaviest elements that can be made through a self-sustaining fusion reaction. This detail is responsible for one of the most spectacular and important phenomena in the cosmos: Stars are the universe's element factories. Without stars, the atoms that make planets, and the life on at least one of those planets, would not exist. As every good factory manager knows, once something is made it must be distributed. Fortunately, the stars, at least some of them, form a marvelous conveyor system that moves their manufactured elements about the galaxy.

A star's effective distribution of its elements depends on its mass, how big it is. The most effective stars are about ten times more massive than our Sun. Like all stars, they form from the condensation of a hydrogen gas cloud. Gravity pulls the cloud together, causing it to grow hot. Upon reaching 10 million Kelvin, the hydrogen ignites, and the heat from nuclear fusion produces an outward pressure, counteracting the pull of gravity. The star then goes about its business of turning hydrogen into helium—when it runs low on hydrogen fuel, fusion slows and the star begins to cool. Once again, gravity dominates, compressing the stellar gas, which is now mostly helium. At the core of this shrinking star, the temperature climbs, eventually surpassing 100 million K, whereupon the helium ignites and fuses to make carbon and other elements. At such extreme temperatures, the star becomes considerably larger than when it was burning only hydrogen, and its expanded surface glows red. These stars are called red supergiants.

A supergiant now goes about its business of making carbon, until it runs low on helium. As before, the star cools, gravity dominates, it collapses, and core temperatures increase. At 500 million degrees, the carbon ignites, fusing with any remaining helium to produce oxygen.

The process repeats—fuel depletion, core collapse, nuclear ignition— and successively heavier atomic nuclei are produced.

Eventually, iron and nickel accumulate in the core and fusion ceases. Temperatures plunge, robbing the star of the outward pressure that opposes gravity. In less than a second, the star collapses, the core temperature rises to 100 billion degrees, but the nuclear fires cannot be reignited. The compression forces nuclei into contact, some of which are inside the repulsive barrier and fuse to make heavier elements like molybdenum, lead, and uranium. However, these fusion products have greater total energies than the nuclei from which they were formed, which could only come from cooling the super-hot stellar core. When this energy is exhausted, the star is in its death throes. Nuclei are very close together and are subject to tremendous electrostatic repulsion. With no available energy to drive fusion, the star's core recoils in an explosive supernova, ejecting the elements built one upon another over a stellar lifetime across the vast expanses of space.

The "new" atoms manufactured and distributed by the supernova now mix with the ancient hydrogen and helium, forming an atomic cloud containing all the hundred or so known elements. Like the first hydrogen clouds, this one will cool until gravity takes over and compresses the elemental soup to form a star. If large enough, it will take up where the last supernova left off—manufacturing more atoms to distribute through the galaxy. Over 10 billion years, galaxies have become rich with atoms of different kinds. Unlike hydrogen and helium, which were formed in the era immediately following the Big Bang, these atoms could combine in a nearly infinite variety, giving rise to the many subtle valleys of the energy ride's Molecular Downs. Here is yet another place where the universe stores energy as it staves off heat death.

8

THE AGE OF MOLECULES

Dust is everywhere, under beds, on picture frames and window-sills. From dust cloths to HEPA filters and electrostatic precipitators, we have developed technologies that do nothing more than rid our homes of this menace. But dust is not confined to living and work spaces. It travels the world. Every year, windblown particles from the Sahara make their way across the Atlantic to wreak havoc on those of us with allergies. Looking beyond the atmosphere, dust plays a major role in the zodiacal lights, the spokes of Saturn's B ring, as well as the diffuse planetary rings about Jupiter, Uranus, and Neptune. And still dust does not become less common as we leave the solar system and venture into the interstellar realm. Fifteen hundred light-years away, in the constellation Orion, the oft-photographed Horsehead Nebula is composed of thick, dark dust. In short, our universe is a dusty place.

Though abundant, dust is a comparative newcomer to the cosmos, and an essential one at that. Dust builds planets. And it is from elements like silicon, oxygen, and carbon that dust is made. It took 10 billion years of fusion before the concentration of these elements was sufficient to make large numbers of the little dust embryos called molecules.

Recall that negatively charged electrons are attracted to positively charged nuclei through electrostatic force. This is the same force pushing nuclei apart; it is also the same force making socks, fresh from the dryer, cling together, and the hair on your head bush out like a porcupine's needles. Things that are oppositely charged—the socks—attract, while like charges—your hair—repel. In the same way, an electron, even one orbiting a single nucleus, is attracted to neighboring nuclei but repelled by their electrons. An electron needs an orbital with attractive forces more powerful than the repulsive electrostatic forces in order to form a molecule. These orbitals must allow electrons distance from one another and encourage "up close and personal" activity with nuclei. Sometimes the orbitals are a wrong match for the electrons and don't permit them to avoid one another, but on numerous occasions, the converse is true. Then nuclei of the right kind are perfectly positioned and a molecule is formed.

Just as the energies of elemental nuclei vary, so do molecular energies. And just as nuclear reactions can produce energy, so can chemical reactions. The crucial difference is the amount of energy liberated: Nuclear reactions produce many times more energy than do chemical reactions—a fact made all too real by the "yield numbers" attached to nuclear weapons.

The energy released by a nuclear weapon, or its yield, is given in units of tons, where one ton is the energy released by the explosion of 2,000 pounds of TNT—a chemical reaction. For some perspective, the large "bunker-busting" bombs used in the Iraq war released about as much energy as two tons of TNT. The yield of a typical nuclear weapon in the U.S. arsenal, which realizes energy through the fusion of hydrogen to helium, is rated at one megaton, that is, one million tons of TNT. The largest nuclear weapon ever detonated was the 50-megaton monster built by the Soviet Union in the early 1960s. This single bomb released as much energy as 25 million bunker busters.

For most of us, the true substance behind these numbers is inconceivable. So instead of tons, think of yield in terms of trainloads of TNT. Virtually everyone has sat at a railroad crossing counting the cars of a slow-moving train. Trains hauling coal are particularly common in Colorado, where I live. Behind the ganged engines of these trains trail a hundred or more 100-ton coal hoppers. Five thousand such trains filled with TNT instead of coal would pack the same energy as the Soviet's giant nuclear weapon. The yield of the roughly one-meter-tall H-bombs currently stockpiled by the United States is equivalent to one hundred trainloads of TNT.

Gram for gram, nuclear reactions deliver about a million times more energy than do their chemical counterparts. Nuclear reactions are like the huge dynamos turning beneath Hoover Dam; though generating the energy needed by millions of homes, the energy is not particularly useful until reduced to the many manageable packets that run computers, TVs, and blenders. Molecular reactions serve the same function, reducing and dividing the energy of fusion into smaller pieces, powering the universe's subtle phenomena.

Chemicals that could be subtly manipulated to release hidden energy have always enticed me. The prospect that I could control this energy, summoning it at my will, was enthralling when I was young. I often thought about my first bomb—the incredible flame-spitting gas can—and by my early teens, I had figured out why it failed to take out the garage.

Gasoline, by itself, is not an explosive. It's when gasoline vapor is mixed with oxygen that things get interesting. The molecules of gasoline combine with those of oxygen to produce water and carbon dioxide. The potential energy of the products is less than that of the beginning molecules—the reactants—which means there is some unaccounted-for energy. A small part can be found in the photons of the explosive flash, but the majority is locked up in the kinetic energy of the fast-moving product molecules, which are traveling many times faster than average—many times faster than the speed of sound, in fact, which is why a shock wave accompanies an explosion.

The problem with my first bomb was simple. Only the gasoline vapors in the top of the can were mixed with oxygen, and when ignited these exploded, consuming nearly all of the oxygen in a single flaming belch. The gases left behind, though very hot, were deprived of plentiful oxygen so the reaction stopped—or at least slowed down. As the can cooled, the pressure dropped, and more oxygen was sucked into the container. This mixed with the hot gasoline vapors and reignited, once again spewing theatrical flames.

A more poignant example of this same phenomenon was seen in the crash of TWA Flight 800, which, on July 17, 1996, exploded in midair killing all 230 people aboard. The National Transportation Safety Board attributed the crash to the explosion of the nearly empty central fuel tank, which had become hot (by some estimates 53°C or 127°F) while sitting on the tarmac at JFK for several hours, air conditioners running. The fuel vapor and air mixture was so explosive that very little heat was required for ignition. It's ironic that, had the central fuel tank been full or nearly so, the space available for air would not have provided sufficient oxygen to support an explosion.

With this insight into explosive "design," I turned my covert bomb-building activities to explosive mixtures. Gunpowder is an explosive mixture of sulfur, carbon, and potassium nitrate. The latter chemical is also called saltpeter and has therapeutic uses, which is why it is sold in drugstores, even to twelve-year-olds. Aluminum foil added to an aqueous solution of sodium hydroxide—lye—reacts to make hydrogen, generating sufficient pressure to inflate balloons. Top one of these off with air, or better yet pure oxygen, and when ignited they make for a satisfying display. But the most intriguing explosive mixture I discovered literally allowed me to leave a mark on the world.

I sometimes wonder if I would have survived my childhood had I grown up with the World Wide Web, which makes information gathering almost effortless. If one is interested in explosive mixtures, just type that phrase into a search engine and bingo—a mouse click away are thousands of chemical reactions that go boom. But in the 1960s a curious mind had to know something about explosive reactions in order to find step-by-step instructions. Since I knew there was something called gunpowder, I could find its entry in the Encyclopedia Britannica, which told me how to make it. There was no general heading for "Explosive Mixtures," however. Accordingly, I learned to decode the secret language of potentially interesting concoctions, and those fruitful phrases jumped off the page. One school day, thumbing through a seventh-grade earth sciences textbook, I saw in the right-hand margin a line drawing of a hand pouring liquid onto a solid. The caption made it clear that the liquid was glycerin and the solid was potassium permanganate. In bolder face type was the sought-after warning "Do not try this without adult supervision." I immediately registered the experiment as promising. Thanks to my friend Rodger, it would not be long before the promise was realized.

Rodger was a lab assistant who had after-school access to the lab, which was located between the biology and earth science classrooms. On a particularly cold day in December, I was helping Rodger clean up after a biology lab having something to do with capillary action—where water in small diameter tubes seems to defy gravity. To better see the water, it had been colored with potassium permanganate, a purple crystal that dissolves readily, turning water the same royal hue.

The lovely solid was just sitting there in a large glass bottle. Glycerin is slippery, and in addition to its use in dishwashing soap, it is a common lubricant in a lab. A drop of glycerin on a glass tube and it will slide smoothly through a rubber stopper. The ingredients for the potentially explosive mixture illustrated in the textbook were in plain sight.

The last bell had rung. The teachers had gone home. It was just Rodger, me, a custodian in some distant part of the building, and a mixture that did something requiring "adult supervision." In no time at all, I filled half of a cup-sized container with the purple crystals and proceeded to drown these with syrupy glycerin: nothing. Rodger and I waited, and still . . . nothing. I became defensive. I had assured Rodger we would produce something more spectacular than thick purple goo, so to prove it I dragged him into the earth science room in search of the book with the warning I'd read. That's when we heard the roar. As we turned, we saw flames, born in the tiny cup, leaping two meters into the air. Exploding grains of potassium permanganate arced outward in colorful, parabolic trajectories. This sustained pyrotechnic display made my gasoline bomb look shabby.

Rodger leapt into action. He opened the third-floor window and, using a pair of sturdy tongs, moved the flaming fountain onto the window ledge. On the busy street below, the small volcano was clearly visible and growing more obvious as it showered sparks into the gathering dusk.

I had been in a situation like this once before. Had Rodger asked, I would have suggested that we let it burn itself out. But he didn't, and in what I thought was a serious lapse of judgment, he doused the flame-spewing container with water. This extinguished the reaction but also broke the cup and dissolved what was left of the potassium permanganate. The resulting liquid ran down the wall of the school, discoloring the beige stucco as it went. Reaching the second-floor window, the moving purple stain dropped to the window ledge, where it froze, causing the trailing liquid to drain into the room the window opened on and onto a desk situated next to the window. Unfortunately, that desk belonged to the principal.

While Rodger sought out the custodian to talk his way into the principal's office and then to work for the better part of half an hour to clean up the mess, I sat in the lab. My sole act of helping involved pouring hot water onto the window ledge where Rodger labored to chip away purple ice. At least the hot water helped warm Rodger's freezing hands.

All next day, I waited for a summons to the office. It did not come that day, nor the next, or the next. I had escaped punishment yet again. Apparently Rodger had done an excellent job removing the outward traces of my curiosity. But the one thing he could not hide was the long purple stain connecting the third floor lab to the second floor office of the principal. The stain persisted, still visible years later when the building was demolished.

Potassium permanganate and glycerin, aluminum and sodium hydroxide, carbon, sulfur, and saltpeter—these are but three reactions drawn from the boundless richness of molecular chemistry. For 10 billion years, stars labored to make elements from three basic ingredients: protons, electrons, and neutrons. With this task well under way, the universe had at its disposal an infinite number of ways to store, release, and move smaller amounts of energy. Before these capabilities could be exploited, however, something new was needed: places to manufacture large quantities of molecules. These places had to be much cooler than stars yet warmer and denser than interstellar space. The solution: planets.

The planets of our solar system owe their existence to a giant star that did its all to convert mass into energy. Some six billion years ago, its nuclear fuel spent, this star culminated with a spectacular display, scattering atomic nuclei and small molecules across light-years of space. In the cold environs of the cavernous cosmos, these atoms and molecules occasionally collided with others, as they were literally pushed about by the light of nearby stars.

Among the atoms populating the cloudy death shroud of this supernova were huge amounts of hydrogen and helium; in fact, these elements accounted for 99 percent of the cloud. There were also quantities of the other "light" elements like carbon, oxygen, and silicon; less abundant were iron and nickel, along with trace amounts of the heavier metallic elements like gold, lead, uranium, and plutonium. Small molecules, formed in the corona of the star before its supernova stage, were also found. Atoms of hydrogen had combined with those of carbon in a four-to-one ratio to make methane; two atoms of hydrogen added to one of oxygen made water; carbon atoms linked with one another to make long chains, sheets, tubes, and balls that were nothing more than soot. Silicon atoms joined with oxygen to make silica—tiny seeds of sand.

Buffeted by stellar winds, these elements and molecules combined in diverse ways. Water molecules adhered to the sand seeds, along with an occasional metal atom and a sooty carbon molecule. Over eons, the supernova's last vestiges were transformed from an imperceptible atmosphere laden with hydrogen and helium into a dusty, smog-filled cloud.

All the while, gravity was at work. It tugged on the thin atmosphere as well as its dusty suspensions, pulling them toward centers of greatest density. As the concentration of gas and particles around these centers grew, gravitational energy was transformed into kinetic energy. Molecules and dust particles collided, friction heated the gas and it began to glow—but still, gravity exerted its will, compressing and further heating the gas. When the temperature climbed to 10 million Kelvin, the nuclear furnace that is our sun kicked on.

When previous stars had flashed to life, they shed their light over the vast emptiness of interstellar space and hydrogen and helium gases were swept away by the onslaught of stellar winds. But for our sun, this was not the case. Its first light fell on a disk of dust and ice 10 billion kilometers (6 billion miles) in diameter. Bit by bit, the particles of this disk accreted. Under the influence of gravity, loose aggregates of sand and ice were created, morphing into planetesimals several kilometers in diameter and further condensing into rocky cores hundreds or even thousands of times larger. These genuine planets sucked planetesimals from the now-thinning accretion disk. The kinetic energy lost in the resulting collisions was transformed into heat, producing a molten mass of rock and metal. The denser molten metals were pulled toward the planet's center, leaving rocky silica to float upon the surface. With the planet's growth came a larger gravitational field. This field attracted and held enough gases to form a new atmosphere.

The precise composition of the atmosphere depended on the distance between the planet and the Sun. At equivalent temperatures, the average kinetic energy of all gas molecules is the same, but because some molecules are lighter than others, they must move faster to produce the same kinetic energy. For example, at room temperature the average hydrogen molecule is buzzing around at 1,770 meters per second (about 4,000 miles per hour). On the other hand, oxygen molecules, being sixteen times heavier, move, on average, at one-fourth this velocity, or 444 meters per second. (These are average velocities. There are molecules moving both faster and slower, as Maxwell revealed. A few will go as fast as 11,000 meters per second.)

Eleven thousand meters per second—or 18,000 miles per hour—is the velocity needed to escape Earth's gravitational field. If molecules have speeds in excess of the escape velocity, they can be lost from the atmosphere. In the course of time, the lighter elements would be the first to be depleted from a planet's atmosphere. This situation would be exacerbated as the temperature increases.

Being the coldest, the rocky planets farthest from the Sun held on to hydrogen and helium as well as all the heavier gases like carbon dioxide, methane, and water. Because there was an abundance of hydrogen, the atmosphere of these planets grew rapidly. Where once there had been planets dense with metal and rock, those planets now had immensely deep atmospheres. One day we would name them Jupiter, Saturn, Uranus, and Neptune.

Mercury, the planet closest to the Sun, was too hot even for heavy gases and never developed a dense atmosphere. The next three planets—Venus, Earth, and Mars—were quite different. Though too hot for hydrogen and helium, the temperature was just right for heavier gas molecules, water and carbon dioxide in particular.

Water and carbon dioxide are two of the three ingredients that power the bulk of human technologies. While we all have experience with water in its various forms—snow, rain, and humidity—carbon dioxide is only slightly less familiar. It makes the bubbles in champagne and puts a head on beer. It is a primary by-product of automobiles and breathing. To us it would be the most trivial and inconsequential of gases, if not for the fact that it is transparent to visible light but absorbs infrared radiation.

Infrared (IR) light is just like visible light, only we can't see it. Some animals can—rattlesnakes, for example. With the aid of specific night-vision goggles, we, too, can "see" IR. Infrared light is emitted by everything. The hotter it is, the more IR it gives off. This is why rattlesnakes are such effective hunters of warm-blooded animals: The snake can follow its prey into a burrow where there is no visible light, but it can still see the IR radiated by its intended meal.

Visible light is emitted by the sun and passes right through a planet's atmosphere. The ground and ocean soak it up. Energy is conserved, so whatever did the sopping begins to heat up and radiate IR. If there are no gases in the atmosphere to absorb IR, the heat is returned to space as photons. This is the case where the atmosphere is composed entirely of oxygen, nitrogen, hydrogen, and/or helium.

However, if there is a greenhouse gas present, such as carbon dioxide, methane, or nitrous oxide, it absorbs the reflected heat and the atmosphere warms. The greater the concentration of greenhouse gases, the hotter the atmosphere will become. This phenomenon is believed to be responsible for global warming.

All of us, but particularly those who live in the high desert, have some experience with greenhouse gases and their potential to cause global warming. Water vapor is a greenhouse gas and the clouds it forms are very effective absorbers of IR radiation. On a clear day, intense solar radiation passes through the comparatively thin atmosphere of high-altitude plateaus, causing the ground to heat. Summertime temperatures can soar. As soon as the sun sets, though, the ground begins to reradiate this heat back into space as IR energy. In a dry climate, there is little to absorb this radiated energy, and nighttime temperatures dive. Even in the hottest parts of summer, when daytime temperatures are in excess of 38°C (100°F), you had better grab a jacket if you plan to be out late in the evening. In Santa Fe, New Mexico—more than 2,500 meters (7,000 feet) above sea level—radiational cooling makes for a downright chilly night. Bring in some cloud cover, or even a little humidity, and it's another story. Now the IR radiated from the hot ground is absorbed, warming the air, and there will be little or no temperature change between day and night.

Atmospheric scientists fear that increasing the concentration of CO_2 will have the same effect on a global scale. However, it will be a long time before Earth undergoes the extreme change experienced by its sister planets. Venus, which is nearly the same size as Earth, has a very thick atmosphere that is 95 percent CO_2. Though the "air" of Mars is nearly pure carbon dioxide, it is so thin that the concentration of CO_2 is all but negligible. On the other hand, Earth's medium-density atmosphere contains small but significant concentrations of CO_2 —a few hundred parts per million.

As might be expected from these numbers, Venus, with its dense atmosphere rich in carbon dioxide, is subject to runaway greenhouse heating. Despite it being "dry" heat, you would still find its 500°C (–850°F) surface most unpleasant. Mars is just the opposite: Its thin atmosphere is unable to mount much greenhouse heating at all. Most of the light that falls on the –55°C (–67°F) Martian surface is reradiated into space. Earth, however, seems not too hot and not too cold, but just right.

Though this is the current trend, four billion years ago these three planets probably had similar atmospheres. Chemical reactions between water, carbon dioxide, and rocks were the vehicles of change.

On Earth, atmospheric CO_2 reacts with rainwater, making a weak solution of carbonic acid. When this weak acid falls on rock, it leaches many different elements, including calcium, which dissolves and collects in oceans. Over time, the concentration of calcium becomes too high to stay in solution and calcium carbonate (chalk) precipitates from the water, settling on the ocean floor as limestone. In other words, on Earth water removes carbon dioxide from the atmosphere, tying it up in rocks. The amount of CO_2 locked in limestone, marble, and other minerals is about the same as that in Venus's atmosphere.

You might think that, given sufficient time, all the carbon dioxide would be trapped in rocks, with none remaining in the air. Deprived of CO_2 and hence greenhouse warming, Earth would cool, oceans would freeze, and our planet would resemble a massive snowball. Indeed, some scientists believe Earth has passed through such "snowball" phases already. But the atmosphere does contain carbon dioxide, which means that nature found a way to get it out of the rocks and back into the air.

Remember, Earth had been left in a molten state by the impacts it suffered while growing from a large hunk of rock and metal into a genuine planet. As the number of impacts diminished, the surface of Earth cooled, solidified, and formed a crust "floating" on the denser hot mantle and core. There, heavy elements created as the last act of a dying star accumulated, particularly thorium and uranium. These transform mass into energy, most of which turns up as heat that keeps Earth's innards gently percolating. Some of this energy is converted into motion, shifting Earth's crust.

Picture a broken hard-boiled egg with cracks separating pristine sections of shell. In the same way, cracks appear in Earth's crust, and the contiguous regions are called tectonic plates. Using the energy in Earth's core, these plates move, smashing into each other along one boundary while pulling apart along another. When tectonic plates collide, one dives beneath the other in a process called subduction. The plate above is pushed upward, building mountains; the one below is carried into the hot mantle. Along the subduction boundary, pressures are tremendous and friction causes temperatures to climb. In these extreme conditions, limestone is first transformed to marble, but at higher temperatures even marble yields. The trapped carbon dioxide is released but remains dissolved in the molten rock. With continued subduction, more limestone releases its long-held gas until the weight of the crust above can no longer accommodate the dissolved gas. Then, like a shaken bottle of champagne, a volcano erupts as the crust pops its cork. Carbon dioxide returns to the atmosphere with a violent explosion and brings with it molten rock, called magma.

In an endless sequence—the carbon cycle—carbon dioxide is rinsed from the atmosphere by rain. The acid rain leaches calcium from rocks and carries it to the oceans, where the capacity to hold calcium is limited. When this limit is exceeded, calcium carbonate drops out of the ocean water, forming limestone and marble. Through subduction, these carbon-bearing rocks are carried into the mantle and release carbon dioxide, returning it to the atmosphere via volcanoes.

On our neighboring planets, carbon doesn't cycle. It is the first part of the sequence that is broken on Venus, where there is no water to remove CO_2 from the atmosphere. Billions of years ago, a rainy day on Venus may have been commonplace, but not anymore. Unlike

Earth, where water vapor is trapped as clouds at comparatively low altitudes, water molecules rise high into the Venusian atmosphere. There, intense ultraviolet radiation splits water molecules into oxygen and hydrogen. The lightweight hydrogen escapes, while oxygen reacts with sulfur to form poisonous sulfur dioxide. If ancient rains once ensnared CO_2 in Venusian rocks, it has long since been returned to the atmosphere by sulfur-spewing volcanoes.

Mars is missing the volcano part of the carbon cycle. At one time, rivers ran and volcanoes erupted across the Red Planet. But Mars is smaller than either Venus or Earth, so it cooled more rapidly. Carbon dioxide accumulated in the Martian rocks, and eventually the great continents drifting on its molten core were frozen in place by a planet grown cold and solid. The volcanoes that once spoke of inner life grew silent, and where water once moved, permafrost collected.

When compared to our neighbors, Earth appears to have a built-in thermostat. If it gets too warm, more water evaporates from the oceans, which falls as rain or snow. The added precipitation removes CO_2 from the atmosphere, moderating greenhouse heating and cooling the planet. If the planet cools too much, less water evaporates, precipitation diminishes, and CO_2 collects to warm the planet. Through the carbon cycle, the temperature remains just right, neither so cold that all water freezes, nor so hot that it boils; it is the perfect temperature for liquid water.

9

FROM LIFE TO FOSSIL FUELS

Without liquid water, there would be no life. No fossil fuels. No us to exploit those fuels. Our story would come to a screeching halt. But the solar system, and I suspect the cosmos as a whole, is full of water—most of it tied up in icy bodies orbiting the sun beyond the edge of our planetary system and extending halfway to the nearest star, Alpha Centauri. The most distant objects make up the Oort cloud and are a source of comets. Some of the inhabitants of this grand cloud (which may have a combined mass greater than Jupiter) are interstellar travelers, circling our closest neighboring stars before returning to swing past the sun.

Scientists believe water is an essential—and exceptional—ingredient of life because of its unique interaction with other molecules. Ammonia and hydrogen sulfide are the two molecules most similar to water, but they cross the finish line at a distant second and third. (Anhydrous ammonia is not the familiar household cleaning fluid, which is primarily water. Pure ammonia is a potentially toxic fertilizer used in commercial agriculture.)

Hydrogen sulfide is said to smell like rotten eggs (though I have never smelled rotten eggs, and I'm willing to bet this is true for most of us, so maybe it's that rotten eggs smell like hydrogen sulfide). This stinky molecule is released from volcanoes, geysers, and hot springs, its signature odor recognizable to visitors of Yellowstone National Park or to the Kilauea volcano in Hawaii. Hydrogen sulfide loosed from Vesuvius in 79 BCE, along with other deadly gases, likely hastened the deaths of the citizenry of Pompeii and Herculaneum.

What makes the water we bathe in, play in, and drink similar to these toxic substances? Some materials dissolve readily in these liquids while others don't. It is the *extent* to which molecules interact with liquid water that make it an exceptional molecule. Think of all the substances that dissolve in water, apparently without limit. Sugar continues to dissolve long after the solution becomes syrupy thick; and though the limit for salt is reached a little earlier, saturated salt water is still noticeably denser than pure water. These are said to be *hydrophilic* molecules, or water-loving. There are also molecules that strongly resist dissolution, or "fear" water, and these are called *hydrophobic*. A common hydrophobic molecule is vegetable oil. Take Italian salad dressing, which is composed chiefly of vinegar, water, and oil. The concoction must be shaken before it's poured on a salad, but even then the oil does not dissolve; rather, it forms an emulsion—small globules suspended in the water-and-vinegar solution. In a minute or two, the globules make their way to the top of the dressing where they combine as a layer of floating oil.

This differentiation leads to a remarkable and oft-used chemical rule of thumb: Like dissolves like. For example, hydrophobic gasoline is insoluble in water, but hydrophobic salad oil is soluble in gasoline (though there is no readily apparent use for the resulting mixture). Skin is largely hydrophobic, which is why we don't dissolve while swimming, surfing, or showering.

While we need not worry about dissolving in a spring rain or while taking a dip in a swimming pool, some common molecules, or combinations thereof, should be avoided. I learned about one such combination at age seven, when I was "helping" paint the exterior of our garage—the same one I had attempted to blow up the year before. At the time, most paints, particularly those intended for exterior use, were oil based, and the only way to clean the inevitable splatter from a painter's hands and arms was to employ a hydrophobic solvent. Generally, a turpentine-soaked rag is an effective way to remove paint from the skin. The turpentine "molecules," though hydrophobic, evaporate quickly, leaving few molecules behind to dissolve into the skin. However, just a little bit of water applied on top of the turpentine will trap these hydrophobic molecules between the skin and the layer of surface water. In order to get away from the water they fear, the hydrophobic molecules diffuse deep into the skin's cell membranes, causing swelling and breakage. The sensation is similar to a painful sunburn.

How do I know? Well, after a few hours of garage painting, I was covered with paint. Not content with a rag, and having paint all over my body, I added about a half-gallon of turpentine to a bathtub full of water and climbed right in. At first everything went as planned—the paint came right off—but in a few minutes my skin began to tingle, then heat, and finally to burn. Every square inch below my neck felt as if it were on fire. Only after several hours did the pain fully subside.

Had I been more of a chemist at the time, I would have reached for the antidote as soon as the tingling and prickling began. What I needed that afternoon was another class of molecules altogether, a class simultaneously hydrophobic and hydrophilic, called *amphipathic*.

Generally longish and with a structure reminiscent of a tadpole, an amphipathic molecule has a hydrophobic "tail" and a hydrophilic "head." These molecules live in a world of existential ambiguity. In water, they float about on the surface, hydrophilic head down, hydrophobic tail up and out of the water—much like a duck feeding on a lake's shallow vegetation. In a hydrophobic solvent (vegetable oil) the situation is reversed: tail is down, head up and out of the oil. Only at the boundary between hydrophilic and hydrophobic liquids are amphipathic molecules at home. Between grease and water, for example, the fearful tails can bury themselves deep within a microscopic grease particle, while the surface plays host to water-loving heads. This particle is transformed from hydrophobic to hydrophilic, and now the whole collection of grease and amphipathic molecules will dissolve in water. Sound familiar? A bar of bath soap is amphipathic. In fact, it's soap *because* it is amphipathic.

If used promptly, a bar of Zest or a squirt of Joy would have robbed the turpentine of its lasting sting. The soap's hydrophobic tails would have sought out turpentine droplets, dissolving in them and encasing each in a shell of hydrophilic heads. Transformed from water fearing to water loving, the turpentine/soap combo would have easily dissolved in the abundant bathwater.

While amphipathic molecules do help keep us clean, they serve a more vital function: forming the test tubes in which nature conducts life-giving experiments. When Earth was still an infant and just a few hundred million years old, great rocky plates floated on a core of molten rock and liquid metal. All about the edges of theses plates volcanoes spewed carbon dioxide, hydrogen sulfide, water vapor, and other gases into an atmosphere already rich in carbon dioxide, hydrogen, nitrogen, and, perhaps, methane. A shallow sea only a few hundred meters deep formed as outgassed volcanic steam condensed and fell as torrential rains. Carbon dioxide, washed from the ancient atmosphere, reappeared as limestone on the ocean floor. In this moist environment rich with life's elements (carbon, nitrogen, and oxygen), and with energy supplied from internal heat, ultraviolet radiation, or lightning, molecules of all sorts formed—including those molecules essential to life: amino acids, sugars, nucleic acids, and lipids.[4]

Lipids are amphipathic molecules from which the membranes surrounding all cells (including skin) are built.* In an aqueous environment, these molecules can spontaneously organize themselves into bilayers, where hydrophobic tails huddle together and isolate themselves from water by hiding behind a dense layer of tightly packed hydrophilic heads. The rudiments of a cell membrane are created when these bilayers become closed shells called *vesicles,* sandwiching their water-fearing insides between two slices of hydrophilic "bread." The stability of vesicles, along with simpler structures called *micelles* (which result from the closure of a single lipid layer), is governed by the interaction of the amphipathic membrane with the liquid in which it is suspended. And this is why water is critical to life: When the intensity of the hydrophobic and hydrophilic interactions are extreme, vesicles and micelles are stable and can survive for long periods. The menagerie of amphipathic molecules that spontaneously form vesicles in liquid ammonia or hydrogen sulfide is extremely restricted, and even then, they are fragile structures.

Due to their stability and diversity, some scientists believe that vesicles and micelles populated ponds and lakes four billion years ago. Each little container consumed a different combination of molecules as it closed upon itself. Some may have ingested tiny mineral grains that are known to catalyze organic reactions. Some may have encapsulated dye molecules that could absorb sunlight. Still others formed around

---

*This is why I said skin was "mostly hydrophobic." To be more precise about my turpentine bath, molecules from the solvent dissolved in the hydrophobic part of the skin-cell membranes.

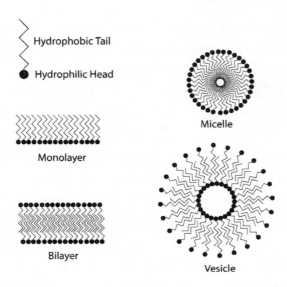

Hydrophobic Tail

Hydrophilic Head

Monolayer

Bilayer

Micelle

Vesicle

Figure 9.1 Amphipathic molecules in water

smaller micelles, born months or even weeks earlier when water chemistry was dramatically different. And though each was chemically unique, the laws of thermodynamics bound them all.

The concentration of chemicals in these primordial ponds was in constant flux. Evaporation lowered water levels and concentrated dissolved molecules; rainfall and runoff had the opposite effect; and inside vesicles, chemical concentrations remained virtually unchanged. The lipid containers responded to these differences by expanding and contracting like bellows. Vesicles may have used these concentration variations as their energy source, or they could have derived energy from absorbed sunlight, initiating catalytic reactions, or by converting dye molecules into a more energetic form.

Regardless of the source, some of the vesicles used this energy to grow by incorporating new lipid molecules into an expanding bilayer. With growth came a kind of reproduction. When a vesicle becomes large and floppy it can collapse on itself—think of filling or deflating a hot-air balloon—then reappear as two or more vesicles that, though smaller, are plump. In turn, each could then grow and divide.

The chemical processes necessary for life were now in place, and vesicles competed with other vesicles for energy and raw material (specific lipids). The most efficient at harnessing energy and using it to do work—grow and divide—proliferated at the expense of the less fit. Over millions of years, a lipid world evolved into a cellular world, exquisitely structured to convert energy into work, thereby slowing the production of entropy. The last vestige of that earlier time is the lipid membrane surrounding all cells.

Even with the advent of living cells, life was restricted to regions where an oxidizer (one component of a reactive mixture) occurred in abundance, perhaps around thermal vents where sulfur is plentiful. The earliest cells persisted by happening upon or maybe even hunting down energy-rich molecules, and harnessing the energy released as these molecules reacted with sulfur. Then something wonderful happened: Inside a bacterium, a tiny change in the structure of one molecule allowed it to use the energy of sunlight to wrestle hydrogen and oxygen atoms from water. Oxygen was poisonous to early life. A much more powerful oxidizer than sulfur, it can strip electrons and break the bonds of almost anything around it—lipids, sugars, amino acids, you name it. Previous cells attempting this trick were abruptly destroyed. In this case, these bacteria had found a way to protect themselves long enough to remove oxygen as a waste product. The hydrogen left behind reacted with carbon dioxide to make sugars and other high-energy organic molecules that, in turn, were used for growth and reproduction.

These photosynthetic cells were free to wander the world. Maybe *wander* is too strong a term; however, they were certainly able to ride the tides and currents of ancient seas before settling down. Now food could be found wherever and whenever the sun shined, and predatory cells, needing to remain close to their oxidizing source, could not follow. Three and a half billion years ago, blue-green photosynthetic bacteria (cyanobacteria) lived in a relatively uninhabited Eden. As they multiplied and took dominion over the planet, carbon dioxide and water were slowly converted into organic molecules and oxygen.

Initially, the resulting oxygen was rapidly consumed as it reacted with iron and other metals. The result was common rust. Two billion years ago, this rust caused Earth's soil and many of its rocks to develop a distinctive red hue. While pure iron remained, oxygen was prevented from reaching the atmosphere. When this depository was exhausted, oxygen bubbled to ocean surfaces. There oxygen reacted with any methane still in the primeval air. And still it came. In less than three billion years, cyanobacteria replaced nearly all the carbon dioxide in the ancient nitrogen atmosphere with oxygen; as a sign of this conversion, the sky turned the lovely blue we associate with our living planet today. Along with this new atmosphere, a gate had opened and a serpent entered Eden.

The cyanobacteria had no defense against predators, but they really didn't need one. They had harvested the energy of sunlight for eons and stored it as life's complex molecules. Like the gasoline in my bomb, the "gas" molecles were only one-half of an explosive mixture. To liberate the energy of photosynthesis stored in the molecules, oxygen was required, and there just wasn't enough of it in primordial seas to initiate combustion. But every year photosynthetic bacteria pumped more and more oxygen into the seas and air, and by 700 million years ago, near the end of the Proterozoic era, the atmosphere and oceans had undergone significant alterations and were now loaded with bacterially generated oxygen.

To survive in this reactive environment, cells needed molecules that would tie up oxygen long enough to get it out of their system before it could do extensive damage. The cyanobacteria had already equipped themselves with an antioxidant (like vitamins A and C) called *catalase* to serve this purpose. In the same way, other cells began to abide oxygen. Among these were predatory anaerobic bacteria, which had up until then made their relatively modest living in oxygen-free environments around deep thermal vents. Now equipped with an oxygen-tolerant metabolism, they ventured into a world of cyanobacteria, which were packed with combustible molecules and surrounded by the oxygen required to support this combustion. In a process called respiration, predatory cells harnessed energy by literally burning their prey. Oxygen from the atmosphere was taken inside the cell where it "combusted" ingested photosynthetic bacteria, reducing them to carbon dioxide, water, and, of course, energy.

After ruling the world for billions of years, cyanobacteria were under attack by life-forms that were no longer intimidated by oxygen. Indeed, these life-forms used oxygen against the cyanobacteria. Some photosynthetic bacteria capitalized on their ability to harness energy from the sun, negotiating deals with living things not similarly equipped. Six hundred million years ago, the bacteria took up residence in these cells and provided their hosts with energy in return for a safer place to live. Plants were born, their light trapping chloroplasts—the relics of cyanobacteria. Some of a plant's trapped energy wound up as heat, and hence entropy; the remainder was used to make, among other things, sugar. In turn, the sugar could be polymerized into two forms—starch and cellulose. Cellulose was useful as a rigid support system as plants moved onto land, roughly 400 million years ago.

Much of the sugar not transformed to cellulose became starch, an attractive energy source for respiring organisms undergoing dramatic change. There is little doubt that animals were feeding on red and green algae 550 million years ago. Fish fossils have been dated to 500 million years ago. And by the end of the Carboniferous period (300 million years ago), predatory dragonflies with wingspans equal to those of modern seagulls hunted smaller insects living among giant tree ferns.

The coevolution of respiring and photosynthetic organisms was necessary if either was to endure. Whereas photosynthesis uses the energy of light to make complex organic molecules and oxygen, respiration returns the products of photosynthesis to starting materials—carbon dioxide and water. The inhabitants of a purely photosynthetic world would starve when carbon dioxide was eventually used up. On the other hand, respiration could not be sustained once oxygen was totally converted to carbon dioxide and water. The two processes settled into a rough equilibrium.

Through respiration, approximately 99.99 percent of organic matter is returned to the atmosphere; the remaining 0.01 percent is buried, preventing it from recombining with oxygen. Over several billion years, this amounts to a lot of buried organic material and an impressive amount of free oxygen. By reasonable estimates, Earth's crust contains 26,000 times more organic carbon than the entire biosphere. If all of this buried organic matter were available to us as fossil fuels, well, oil would be as common as the air we breathe. However, only a tiny fraction of a tiny fraction of once-living matter is transformed into viable sources of energy—proof that the conditions needed to turn organic matter into oil, gas, and coal are rare.

Organic matter begins its transformation to fossil fuel when buried deep in an oxygen-free, or anoxic, environment. Anoxic conditions are typically associated with specific circulation patterns in marine basins—like the Red and Black Seas—where a constant rain of dead plants, animals, and fecal matter carries organic nutrients to the sea floor. There, bacteria mine energy by using dissolved oxygen to support respiration. Bacterial decay requires a constant supply of dissolved gases from the surface and therefore requires a specific circulation pattern.

When evaporation losses from sun and wind exceed freshwater input from rain and runoff, surface water becomes more saline and sinks, taking freshly dissolved oxygen with it. However, when freshwater flow to the sea exceeds evaporation losses, surface water is less saline and "floats" on the denser deep water. If the latter conditions persist, deep water becomes depleted of oxygen, creating a dead zone free of aerobic life—including bacteria. Here, once-living plants, animals, and microbes become encased in mud and sediment and are sealed off from the bacteria that may return when the weather changes. (This is how marine fossils are formed.) The Black Sea is the closest modern example: A thousand years ago, the sea floor was fully depleted of oxygen. Its original sediment contains more than 10 percent organic matter.

The next step to making oil requires heat. Organic matter must be baked in nature's oven by burying it in the oil window—that part of the Earth 2,200 to 4,500 meters (7,500 to 15,000 feet) below the surface. At the top of this window, the 80°C temperature is sufficient to start cracking organic biopolymers into shorter molecules containing the five to twenty carbon atoms characterizing the black goo we call crude oil.* At greater depths, cracking progresses to still shorter molecules until, at 4,500 meters, all that is left of the principal deposit is gaseous methane—a molecule made from one carbon and four hydrogen atoms.

After converting to a liquid, oil floats upward, insoluble and lighter than water. Over 90 percent of all oil seeps to the surface; the remaining 10 percent is trapped underground. If it is to remain there, it needs a proper seal. Over geological time, even a tiny seep is sufficient to drain an oil reservoir: A leak of one drop per second will empty a billion-barrel oil field in about 100 million years (not so long in oil years). The Roper Superbasin in northern Australia holds oil that was cooked over a billion years ago. Keeping the lid on an oil basin, like

---

*Think of organic matter as long chains of carbon atoms tangled together. As temperatures and pressures increase, these crack into shorter chains, and those containing five to twenty carbon atoms are what constitute liquid crude oil.

this one, is the job of a caprock, which is generally made from very fine-grained stones.

Though the oil is there—underground—it is trapped in tiny pores that cannot be tapped unless the ensnaring rock beneath the caprock is also porous. To flow up and out, these pores must be connected, making the stone permeable, like sandstone. (While some stones are porous, they are not necessarily permeable. Lava rock, or pumice, is a good example.)

So there we have the recipe for oil: organic matter deprived of oxygen, roasted slowly, and then sealed within porous and permeable rock. Any flaws and you may get absolutely nothing. That's why Earth is abundant in buried organic matter, but so little of it makes high-grade crude oil like that from the Middle East, which is as good as it gets. Middle Eastern oil is cooked smack in the middle of the oil window, suspended in rock with excellent porosity and permeability, then sealed in place by good caprock. Runners-up in the crude oil contest are West Texas and the North Sea.

If you're willing to compromise perfection, however, there are alternatives to high-grade crude oil. Heavy oil and tar sands result when crude oil is not sealed properly in place. In that case oil can migrate too close to the surface, where gases and lighter hydrocarbons evaporate. In this near-surface environment, the remaining crude is subject to the degrading effects of bacteria: When oxygen is available, bacteria eat hydrocarbons, and together these form a viscous substance known as heavy oil or, in extreme cases, tar. These viscous fluids are difficult at best to pump, and in some cases they must be recovered through surface mining. The estimated 1,200 billion–barrel reserve located on the northern bank of the Orinoco River in Venezuela is the largest-known source of heavy oil. Potentially as bountiful are the Athabasca tar sands in Alberta, Canada.

Forget the caprock and you get viscous product; forget to cook organic matter and you'll end up with oil shale. The world's largest source of oil shale lay in the Green River Basin straddling Colorado, Utah, and Wyoming.* Snowmelt from this region is now drained by the Green River and its system of tributaries, but 40 million years ago the runoff was trapped in an inland sea ten times larger than Utah's Great Salt Lake. From the salts and minerals it left behind, we know that this body of water was characterized by an unusual chemistry that would have left its bottom oxygen-starved—the first of nature's requirements for oil. But that's it. The organic matter at the bottom of this sea was never buried into the oil window and remains visible today as dark bands painting the dry buttes and mesas that are so characteristic of the desert Southwest.

Though it was never cooked, the oil shale is basted and ready to pop into the oven—then we'd have black gold. However, the very best shale delivers but one barrel of oil per ton, which means that for every barrel produced there would be about 1,800 pounds of hot crushed rock to dispose of. These processing costs dim (though don't destroy) the luster of this black gold.

Today coal is the world's most abundant oil substitute. Known reserves (the largest of which are in the United States and the former Soviet Union) are sufficient to meet current demand for hundreds of years. Unfortunately, burning coal has a long list of detrimental environmental consequences, including but not limited to atmospheric carbon dioxide, smog, acid rain, and mercury pollution.

For much of this coal, we are indebted to the many bogs and swamps that formed shortly after flora expanded onto land 400 million years ago during the Carboniferous period. In these marshes, plant growth exceeds decay, causing a mat of organic material called peat to accumulate on the soggy soil. As it grows deeper, the dead plant material on the bottom is cut off from oxygen. If this material is subse-

---

*Estimates put the potential oil from this source at between 500 and 1,500 billion barrels. Compare this with the estimated 400 billion barrels remaining in Middle Eastern oil fields.

quently buried—as opposed to being lost to erosion—it is well on its way to becoming coal.

In coastal Carboniferous peat bogs, changing sea levels assisted burial. This period was marked by repeated expansions and contractions of polar ice sheets. With each contraction, ocean levels rose and coastal bogs flooded; with every expansion, the water receded, leaving behind a thick layer of sediment and the potential for a new marsh to form upon the grave of the old. After dozens of events, great quantities of plant material were trapped under deep layers of this sediment.

It just so happens that the temperatures required to cook coal beds are those very temperatures found in the oil window. Proceeding downward, the carbon content of the resulting coal increases, from low-carbon lignite at the top to high-carbon (90 percent) anthracite at the bottom of the oil window. The most common commercial product is bituminous coal, which was cooked in the middle of the oil window.

Natural gas is the final alternative. Depending on its formation temperature, natural gas—methane, mostly—exists in all forms of oil, to varying degrees. When cooked near the top of the oil window, little methane is produced; at the bottom, that's all you get. Natural gas is a relatively clean fossil fuel, producing less carbon dioxide and fewer pollutants than either coal or oil. The only glitch is that no one knows for sure how much natural gas is out there.

When we look around the solar system, we find a good deal of methane. Jupiter's atmosphere, which is mostly hydrogen and helium, is about 0.1 percent methane. This has led some to speculate that during its formation Earth also may have trapped methane—gas that is *not* the product of decomposing organic matter. If this is true, then there might be vast undiscovered supplies of natural gas. The argument goes that we haven't found these reserves because we have been looking for gas in all the wrong places, the same places we look for oil, instead of the deep rocks where it would have been encased billions of years ago.

To date, there is no substantive evidence to support this speculation—but that doesn't mean it's not possible.

A fact that is not in dispute is that there are immense quantities of natural gas caught in molecular cages made of water called hydrates, also known as clathrates. When pressures and temperatures are just right, water molecules can freeze into these structures that, if built with Tinkertoys, would look like a network of linked cages. Each cage is barely big enough to hold a single gas molecule. Under the Arctic permafrost and in the deep water on the outer parts of the continental shelf, there are enormous clathrate fields that have ensnared natural gas in their molecular cages. Inside these methane hydrates there may be one hundred times more natural gas than is held in the world's known conventional gas reserves. Like so many of our energy resources, we have yet to explore technologies to exploit clathrates.

Fossil fuels would have gone unrecognized as the little packages of energy they are if not for evolutionary changes forced on an unremarkable primate by a drying landscape. These changes would lead this creature to develop an imagination and seek a way to give forms to the things it imagined. The species' future was inextricably linked to energy and its availability.

10

Human Evolution and the Origins of Imagination

Three million years ago, something happened to launch a medium-sized primate species into imaginative, energy-dependent human beings. What this something was is one of the monster questions of science. Paleoanthropologists continue to debate the question of human evolution—when and how did we become human?

As befits a monster problem, there are almost as many theories explaining our origins as there are paleoanthropologists. One posits that we have evolved from aquatic hominid ancestors. Another attributes what we are now to our female progenitors, who fashioned the first tools to carry infants as they gathered food while males were off hunting, often unsuccessfully. (Of course a corresponding theory gives primary credit for human evolution to males.) If history is any indicator, it is unlikely that any of these theories is correct. Monster problems are almost never solved all at once; rather, the answer grows by accretion, as solutions are found to small parts of the larger question. So we are left with many plausible scenarios, none of which can be confirmed experimentally.

When scientists evaluate equally plausible hypotheses that cannot be tested, we generally come down on the side of the most elegant solution. By elegant, we mean the theory that requires the fewest assumptions, the simplest solution, also called the principle of parsimony. Einstein implicitly referred to parsimony as "the noblest aim of all theory," which was "to make [the] irreducible elements [of a theory] as simple and as few in number as is possible." The great triumphs of physics are monuments of parsimony: Maxwell's equations, the postulates of quantum mechanics, and the basic rules of special and general relativity are concise enough to be printed on T-shirts.

Since Darwin's time, scientists have recognized that a parsimonious theory of human evolution should somehow link our bipedal locomotion and our large brains. To this, others have attached speech, and I believe that sweaty, hairless, and imaginative should be appended as well. These attributes can be explained with an elegant story that has as its central motif the curious relationship between our ancestors and energy.

ACT I

The first act of our great ancestors' transformation from gentle, arboreal beings into the sweaty, hairless, talkative, bipedal, imaginative, energy-loving creatures we are today is told in full by Steven Stanley in his book, *Children of the Ice Age*. Act I begins in a region now known as the Isthmus of Panama. I say "now known" because 10 million years ago, the continents of North and South America were separated by a vast oceanic gateway called the Central American Seaway. Through it, currents carried the waters of the Pacific into the Atlantic: in effect, returning water to the Atlantic that the hot, dry trade winds of the Sahara had evaporated from its surface and then carried westward to fall as rain on the Pacific.

The Central American Seaway, however, was not built on a solid foundation. It was situated above colliding tectonic plates. The southern plate, upon which South America sits, sought to occupy the same space, at the same time, as the more northerly Panamanian microplate. The laws of nature do not permit this, so something had to give. And what gave was the floor of the seaway, which lifted as one plate plunged beneath the other.

Over millions of years, the seafloor rose, islands were formed, and sediment collected in the shallow waters among them. Together, these obstructions slowed the currents returning water to the Atlantic, stopping them altogether three million years ago, as an archipelago became an isthmus. With the completion of the great land bridge connecting North and South America, the vaporized moisture from the Atlantic that fell as rain into the Pacific could no longer return to its source. This caused the water of the Atlantic north of the isthmus to become more saline and denser than that of its sister oceans.

Though the formation of the isthmus made the North Atlantic only two parts per thousand saltier than the Pacific, this was enough to close the register on a great conveyor of heat to the Arctic. For many millions of years, tropical water was shuttled north on a windblown Atlantic current, bathing Europe in warmth: Tropical plants grew along an extended arm of the Adriatic Sea south of the Alps; cypress swamps fringed the northwestern Mediterranean; and an extinct variety of hippopotamus roamed the riverbeds of England and Wales.

Even as heat was surrendered to land, this current continued northward, spilling into the Arctic Ocean with sufficient warmth to keep it ice-free much of the year. But with the formation of the isthmus, the water carried by this current no longer flowed into the Arctic Ocean; being denser, it sank upon encountering these less saline waters. At great depths, it turned south and then eastward, around Africa and India, then into the northeastern Pacific, where it would rise and return as shallow water to the North Atlantic. So was born the thousand-year cycle of oceanic circulation known as the Global Ocean Conveyor, which continues to this day.

Deprived of a heat source, winter ice did not entirely melt from the Arctic Ocean in the summer months. The persistent ice reflected solar radiation back into space, exacerbating Arctic cooling and initiating a feedback loop that, 2.7 million years ago, resulted in the formation of a northern ice cap. Earth had entered an ice age. As glaciers reshaped northern landscapes and the oceans cooled, less moisture was pumped into the atmosphere, rainfall decreased, and the great rain forests of southern Africa began to recede in favor of savannahs. Living in the woodland margin between the two climatic zones was a primate species named *Australopithecus africanus*, literally "southern ape from Africa."

Slightly smaller than a modern chimp, a female *Australopithecus* is estimated to weigh 30 kilograms (66 pounds); males were 50 percent larger. Though slighter of body, the few adult skulls that have been found suggest that *Australopithecus*'s brain was slightly larger than a chimp's, though not by enough to make them intellectual standouts among Pliocene primates. The wide flat teeth found in these skulls indicate that *Australopithecus* fed primarily on fruits, nuts, and leaves, though, as with chimpanzees, its diet may occasionally have been supplemented with the meat of small animals.

Anatomically, *Australopithecus* was adapted to woodland life, having lived for a hundred thousand generations among a spotty canopy of trees that allowed light to penetrate to the woodland floor where grasses grew and herbivores grazed. With only a partial tree canopy, *Australopithecus* was neither fully arboreal nor terrestrial, descending to the ground to move from tree to tree. Fossils of *Australopithecus*, of which the most famous is Donald Johnson's discovery of Lucy, a three-million-year-old (pre–ice age) skeleton of the northern species *(Australopithecus afarensis)*, reveal a pelvis, knee, and arched foot evolved for bipedal locomotion. The large toe appears to be opposable, a useful trait for tree climbing but one that would have made the creature's gait slow and inefficient. *Australopithecus*'s shoulder, arm bones, wrist, and fingers are characteristic of arboreal primates, such as chimpanzees.

From the fossil evidence, we can imagine *Australopithecus africanus* foraging for fruit, seeds, and leaves, both in and around trees. However, being much too slow to evade danger on the ground, *Australopithecus* could not venture too far from the safety afforded in their boughs. *Dinofelis barlowi,* "terrible cat," was only one of several fearsome ice age predators for which primate was undoubtedly a dietary staple.

Australopithecus's occupation of a spot near the bottom of the Pliocene food chain is supported by more than simply informed speculation. Charles Brain has examined the fossilized remains found in Sterkfontein Cave in South Africa. Among these are large numbers of *Australopithecus* and baboon skulls and jawbones, though bones from below the neck of these species are comparatively rare. Brain argues that this distribution of remains is consistent with modern predation patterns: Carnivores and scavengers crack the long bones of their prey for the marrow, ignoring skulls and jaws.

The scene Brain paints is not of a relaxed species living off the land, but one of a species constantly under the threat of predation by the formidable carnivores roaming the woodland. Despite these pressures, *Australopithecus* survived, almost without change, for more than a million years. The only possible explanation is that they learned to beat a hasty retreat into the nearest tree when danger presented itself.

Two and a half million years ago, the modern ice age was fully under way. Rain forests shrank and the vast deserts of the Sahara and Kalahari expanded; the transitional woodlands at the boundaries of the rain forest contracted. Over millennia, these became sparse as trees grew smaller and groves more separated, which posed two problems for *Australopithecus*. First, the fruits and seeds that nourished *Australopithecus* were no longer provided by a single copse of trees, requiring the primates to wander farther to satisfy daily food requirements. Which brings us to its second problem: carnivores. This was too much for the diminutive hominid, and in an instant of geological time, the Pliocene predators drove *Australopithecus* to extinction.

ACT II

While *Australopithecus* was starting down the road to oblivion, a small group of these primates was taking a different route. We can only guess as to how this band of *Australopithecines* survived for generations as the fearful carnivores of ice age Africa consumed their cousins. However, it is almost certain they became isolated from the rest of their species in a locale that provided protection from nocturnal predators, for night on the ground in Pliocene Africa was even more foreboding than it is today.

It is at night that modern lions and hyenas wreak havoc on the grazing animals of the grasslands. Lions, which sleep during the midday heat, awake at dusk to begin their feeding. By day a scavenger, the hyena becomes a ruthless killer at night. Hunting in packs, both lions and hyenas converge on their prey, confusing the animals, and in their confusion, one or more unfortunate beast will run headlong into the waiting jaws of a pack predator, who will then share the kill with the others of its pride or clan.

The terrors of the modern-day African night are well illustrated in the lore surrounding the building of the Mombasa–Lake Victoria railway in Uganda. Though smallpox, malaria, and tsetse flies killed hundreds, what is remembered most are the lions—especially those that preyed upon the bridge building crew at Tsavo in 1898. Lieutenant-Colonel J. H. Patterson, who was in charge of the crew, records in his book, *The Man Eaters of Tsavo,* his attempts to stalk and kill these lions. Every night for ten months, he would take to a tree with a .303 rifle and a double-barrel shotgun loaded with solid lead slugs. Then he would wait.

> But all in vain, [for] either the lions saw me and then went else-where, or else I was unlucky, for they took man after man from different places without ever once giving me a chance of a shot at them. . . . Once they reached the vicinity of the camps, the roars completely ceased, and we knew then that they were stalking for their prey. Shouts would then pass from camp to camp, "Khabar dar, bhaieon, shaitun ata" (Beware, brothers, the devil is com-ing), but the warning cries would prove of no avail, and sooner or later agonizing shrieks would break the silence and another man would be missing from roll-call next morning.

With a secure retreat, surviving the night was probably easier than getting through the day, when *Australopithecus* had to be concerned with food. All through the early morning, predators laid siege to the borders of the hominid's safe haven, having learned from experience that the defenseless primate would be driven by hunger to the savan-nah. But as they waited, *Australopithecus*'s single ally sapped the ene-mies' will: The sun climbed higher into the tropical sky, its radiant energy beating down on the backs, necks, and heads of the Pliocene carnivores. Threatened with hyperthermia, they eventually retreated to the shade of an isolated acacia tree or shrub thicket to rest until the sun relented. Now, in the midday heat, *Australopithecus* exploited its sole advantage—bipedal stance—and ventured forth in search of food.

At midday, *Australopithecus* took advantage of its own shadow, using head and shoulders as an umbrella. As radiant heating by the sun is directly proportional to the surface area on which it shines, *Australopithecus* was able to tolerate more intense radiation than the carnivores that fed on it. As the day progressed this advantage faded, however, and disappeared altogether by midafternoon, when the sun was halfway between zenith and horizon.

So, forget dinner, forget breakfast, the survival of these few *Australopithecines* came down to their ability to move about on the hot African savannah and find a nutritious lunch. Over years, the ability to "find lunch" shaped their progeny. In each generation there were those better able to perform the task, and thus survive in greater numbers to pass their survival traits on. Among the more important of these traits was the ability to tolerate heat.

The most heat-sensitive organ in a mammal's body is its brain. Changed by a few degrees Celsius—up or down—the brain ceases to function, and the unfortunate animal dies. To avoid this, heat is removed through the circulatory system by blood, which acts as a heat exchanger. For example, blood is cooled as it passes through the many veins of an elephant's flapping ears. This cool blood is then shunted to the heart and back to the elephant's brain, where it picks up heat to be dropped off when, once more, it passes through the veins of the ear. In comparison with their bodies, the ears of Asian elephants are a third as large as those of the African variety, indicative of the less stressful thermal environs of the shaded Asian jungle compared to the wide-open African veld.

If we knew the average ear size for a population of elephants through many generations, we would be able to piece together the history of their thermal surroundings. Unfortunately, ears don't leave fossils. However, as Dean Falk revealed in *Braindance: New Discoveries about Human Origins and Brain Evolution,* the pattern of cranial blood flow in hominids is printed in telltale markings on the bones of the skull. Thus, the story of humankind's evolution is encoded in the fossil record.

Before decoding this fossilization story, we need to understand the requirements placed on cranial blood flow by bipedal locomotion. The act of getting around on two legs increases the demand for blood to the nerves of the spine. On the return trip to the lungs, a network of veins at the back of the neck that surround the spinal cord, called the vertebral plexus, satisfies the demand. When standing upright, it is supplied with blood through a direct connection to the veins of the head and brain, which, as they effect heat dissipation, are of two types: First, there are veins that are wholly contained within the skull. The blood in these *meningeal* and *diploic* veins drains from the braincase through venous sinuses that exit the skull at its base. In turn, these sinuses supply the vertebral plexus. Second, there is a network of small veins on the surface of the skull. These veins communicate directly with the vertebral plexus, as well as penetrating the skull via emissary veins to join with those of the braincase.

Because the emissary veins transport blood across the skull, only these can move heat from the brain to the veins under the skin of the cranium and, ultimately, to the air. Collectively known as the radiator network of veins, it is the equivalent of an elephant's ear. And, like an elephant's ear, if we could chart the variation of the radiator veins in a population of hominids over many generations, we could deduce the changes occurring in their thermal habitat.

In the absence of radiator veins, all the blood supplied to the vertebral plexus flows through the venous sinuses. To accommodate this large blood flow, some of these expand, leaving a dramatic groove on the back of the braincase—a groove that is preserved and easily discernible in fossilized skulls. The radiator veins also leave evidence of the holes through which the emissary veins of the radiator pass into the skull. Though most of these holes are microscopic, a few are large enough to survive fossilization. These are called *emissary foramina,* and their existence proves to be a good indicator of the cooling capacity of the skull's radiator. A fossil skull with deep sinus grooves and no *emissary foramina* is the functional equivalent of a tiny radiator, indicating the biped lived in a shady environment and had little need to shed cranial heat. On the other hand, the large radiators required by bipedal savannah dwellers would be characterized by the absence of a sinus groove and the presence of *emissary foramina.*

Piecing together the story of our ancestors' lives in the sun, therefore, came down to counting holes and measuring grooves, which is exactly what Falk did in every fossil cranium she could find: *Australopithecus africanus, Australopithecus afarensis, Homo habilis, Homo erectus,* ancient and modern *Homo sapiens,* and Neanderthals. And as she counted, a new window on human evolution opened.

The skulls of African apes, which Falk assumed represented our arboreal progenitors, displayed neither deep groves nor emissary foramina. For bipedal *Australopithecus,* the story was more complicated. The northern species, *afarensis,* showed a groove indicative of enlarged venous sinuses but an absence of foramina, while in the south, *africanus* possessed the opposite configuration—a few foramina but no grooves. In the species of *Homo,* the presence of enlarged sinuses fluctuated and declined to very low frequencies in modern humans, while the frequencies of emissary foramina increased. Falk concluded that the whole system of cranial blood flow in the branch of hominids leading to modern *Homo sapiens,* was in the direction of bigger radiators, and at the root of this branch is *Australopithecus africanus.*

Recall that survival of the band of *Australopithecines* on the African veld was congruous to their tolerance of the midday sun. That ability was enhanced in individuals with bigger radiators. Over many generations, a more complex blood flow evolved, allowing the brain to better cool itself. *Australopithecus africanus* was becoming *Homo ha-*

bilis. At the other extreme was *Australopithecus afarensis*—the northern descendants of Lucy—who, without even the rudiments of a brain-cooling system, were doomed to extinction by the changing climate and predators of the ice age.

Homo habilis, for "handy man," made an appearance about 2 million years ago. Perhaps taller and slighter of build, its most distinguishing characteristic was a larger brain, 650–750 cubic centimeters, compared with 450–500 cc for *Australopithecus.* In turn, just 400,000 years later, *Homo habilis* gave rise to *Homo erectus,* which looked nothing like *Australopithecus.* This hominid was tall, perhaps 1.8 meters (six feet), with a cranial capacity of 800 to 1,100 cc, almost 75 percent that of modern humans.

A big radiator will cool a bigger brain, so once the thermal constraints are removed, a brain will naturally grow. But this presupposes that big brains confer dramatic survival advantages, which is not necessarily true. After all, the brain is an expensive organ to maintain—it uses a huge amount of energy. Consequently, it's limited to just the size required to get the job done. And generally, the most demanding part of that job is finding food. So how did early *Homo* catch its prey?

ACT III

Act III turns on an existential crisis: *Australopithecus/Homo habilis* had to consume enough calories in a few hours to sustain it until the following day, when once again the sun forced predators from the savannah. Unfortunately, plant matter is not particularly nutritious, which is why herbivores eat almost all day long. But with only a short lunch break, *Australopithecus/Homo habilis* was forced to abandon its predominately vegetarian diet in favor of more caloric fare: other animals.

Only the most superficial hunting records of early *Homo* exist, and the majority of these pertain to the species upon which they preyed. How they caught these animals is a mystery. In the absence of definitive evidence, it seems reasonable to infer something of early *Homo*'s hunting methods from those of modern-day hunter-gatherers living in similar environments. Take the San (Bushmen) of the Kalahari, for example.

The lives of the San hunters have been skillfully captured in the documentary *The Great Dance: A Hunter's Story,* directed by Craig Foster. The film chronicles !Nqate, Karoha, and Xlhoase who, just as their ancestors did for thousands of years, trek the Kalahari in search of game.* In their recorded hunts, they drive a cheetah off its kill, slaughter and cook porcupines, and remarkably, at the hottest time of day, in the hottest time of year, Karoha chases a healthy kudu to exhaustion and death. While each of these hunts calls for different physical skills, they were all made possible by a common mental ability: locating prey.

Predatory success is tied to the ability to find prey. Most predators rely on an acute sense of smell, hearing, or vision for this purpose. For the San, this would not work. Their long-distance vision is not as acute as that of a wild dog or eagle, for example, and their senses of smell and hearing are grossly inferior to that of almost all major hunting and savaging species. Still, the San are great hunters. They track their prey.

In the sands of the Kalahari, the San deduce the recent history of their surroundings—what animals have come and gone, when they passed, and where they were going. San hunters recognize the tracks of nocturnal animals and insects; if these lie on top of the print of, say, a cheetah, the cat was there the previous day. Conversely, if the cat track obscures the almost imperceptible mark of fallen dewdrops, the cheetah left its print sometime after sunrise and may be close by, perhaps

---

*The San language incorporates nonphonemic clicks that are indicated with punctuation marks, as in !Nqate.

having killed an antelope the San hunters might steal. Windblown sand fades impressions, and this is a veritable clock to judge the time elapsed since a track was made. Tracking is a pleasure and a pastime for the San. Their existence depends on this learned skill, which they teach to their children, grandchildren, and great-grandchildren.

Though the skills of early man could not have been as developed as those of the San, *Australopithecus/Homo habilis* had to perfect rudimentary tracking skills before they could supplement their diet with meat. Unlike other predators, which relied predominately on their senses to locate prey, *Homo* inferred from tracks in the soil where food was to be found. For early *Homo,* finding food became a mentally as well as physically demanding exercise. With its cooling system in place, the brain was free to grow in response to these new demands.

Simply knowing where game was did not guarantee early *Homo* successful hunts. They still had to kill animals that were better runners than our earliest ancestors were chasers—at least over short distances. Luckily, *Homo habilis* was more tolerant of midday sun, which, when combined with the ability to track, provided an efficient hunting strategy.

Early *Homo* used heat as its claws and jaws, capitalizing on a bipedal stance and brain-cooling radiator to remain if not cool at least alive while prey was forced into hyperthermia and, ultimately, death. This remarkably efficient killing tactic did not stop there, for accompanying *Homo habilis*'s evolution came an unparalleled collection of body-cooling adaptations that, when combined with an expanding brain and ability to track, made *Homo erectus* the top predator of the daytime savannah.

Recall that *Homo erectus* was tall and lean, perhaps 1.8 meters (six feet) and 75 kilograms (165 pounds). *Homo habilis* was only four feet tall and weighed 100 pounds. This dramatic evolutionary response did not result from selection for better gathering and scavenging;

rather, it conferred on *erectus* the ability to run down large game, just as Karoha in *The Great Dance* was able to run four relentless hours, chasing a herd of kudu.

Kudu can run at over 80 kph (50 mph) for several minutes, easily outrunning a hunting lion. But after these bursts of speed, their bodies must cool, which on a hot day can take many minutes. On the other hand, an extraordinary human runner can sustain speeds of 20 to 25 kph (12 to 15 mph) for many hours, as recorded in the astonishing scene in Foster's film. Karoha was able to track the herd wherever it sought refuge, keeping the animals running and denying them necessary recovery time. In the end, an exhausted and bewildered kudu stood, unmoving, as Karoha delivered repeated spear thrusts to its vital organs.

This dichotomy is what makes it possible to drive to heat exhaustion an animal that is stronger, faster, and more efficient than the chasing Bushmen. Under normal circumstances, this strategy is fruitless: The speed of the antelope transports it far from the chasing hunter, where it fully recovers before the runner closes the distance. Only on the hottest days, when the prey needs more time to recover, do the runner's physiological adaptations to exercise in the heat shift the advantage in favor of the hunter.

In dry climates, the rapid evaporation of moisture from the skin provides an effective body-cooling mechanism. While humans are not the only animals who sweat, we are the most prolific sweaters. For example, during extreme physical exertion, pigs (contrary to the popular adage) produce no sweat; they have no sweat glands. Horses, however, produce 100 grams of perspiration for each square meter of skin per hour; camels, 250 g/m² hr. And for humans, it's a whopping 500 g/m² hr. This translates into a water loss rate of as much as one to one and a half liters per hour, which, not surprisingly, is the same amount physical trainers recommend consuming during exercise. This evaporation rate is sufficient to dissipate the heat of ten 60-watt incandescent lightbulbs, giving rise to an incredibly effective cooling scheme.

But the heat to evaporate perspiration comes from either the surrounding air or through the skin. Greater body cooling results if most of this heat is supplied by the latter source. This is why being "naked apes" maximizes the cooling efficiency of our sweat. In the absence of hair, each drop of perspiration contacts the skin over a greater surface area and, as it evaporates, takes significantly more heat from the body than from the air.

We enjoy one more body-cooling adaptation: breathing at will and through the mouth. As a quadruped runs, its body works like a bellows, forcing it to take one breath with every stride. As the back extends, the lungs expand, forcing an inhalation; when the back contracts, the lungs compress, causing an associated exhalation. Because nasal breathing (typical of apes) offers too much resistance from narrow nasal passages to support high ventilation rates, physiological adaptations permit us to breathe through the mouth, a more effective means of cooling. As we run (or for that matter, as *Homo erectus* ran), breathing is independent of stride. Without having to run faster, we can take as many breaths as necessary to support the oxygen flow that running demands. Equally important, this respiratory control provided the foundation for breath control, making speech possible.

In the elegant drama of human evolution, all the cooling mechanisms we possess evolved in *Homo erectus*. Unlike so many animals whose evolution was tied to the efficient use of energy, *erectus* took a different road: Rather than selecting offspring that ran faster in order to pull down larger, stronger prey, our ancestors were selected for their ability to stay cool. Fitness was no longer coupled with using energy resourcefully but rather with dissipating the heat generated from energy use. Liberated from these constraints, the ancestors of man were free to evolve a big, energy-consuming brain and an equally hungry gait. Along with these inefficiencies came an advanced brain-cooling radiator and a corresponding system of sweat glands and respiration.

As the curtain drops on Act III, we see a band of hominids, slightly taller than modern humans, running across the hot African veld in pursuit of game. The midday sun glistens on their sweat-covered, hairless bodies. It is unclear precisely what kind of game they are following, for it is hiding. Only the tracks left in the sand betray that it passed here, and it is these the band of hominids follows. Gesticulating to one another as they decipher the meaning of the enigmatic tracks, they quickly change direction, frightening an antelope from its concealment. Hours from now, the chase will end, as the antelope's neurons begin to fail due to an abnormally high brain temperature. On hot days, the hominids are always successful, and *Homo erectus* is the supreme predator of the daytime savannah.

ACT IV

This act of the elegant drama of human evolution recounts *Homo erectus*'s transformation from a brainy animal to an even brainier, imaginative animal. As the act opens, *Homo erectus*'s 1,000 cc brain is capable of making sophisticated stone tools—but only of a design that arose naturally. *Erectus* observed beneficial mistakes made during manufacture, then duplicated these in making the next tool. This brain also appears to have learned to make and control fire, a belief that is supported by 1.6-million-year-old *erectus* campsites discovered in Kenya in the last century. Among the tools and fossilized remains at these sites are lens-shaped burned patches of ancient campfires.

With the discovery of fire making *erectus* was able to migrate from the African continent. Even as the top diurnal predator, *erectus* had to seek shelter from nocturnal equivalents so couldn't wander too far from its nighttime shelter. But with the ability to make fire, protection from savannah predators was guaranteed by constructing a fire-centered circular barricade of thorny brambles, just as the San hunters do today when venturing far into the savannah.

Still, *erectus* was not imaginative. It appeared that, on the way to becoming the top predator of the day, a Faustian bargain had been struck to prevent the *erectus* brain from getting any larger than necessary for 1) following animal tracks; 2) duplicating the beneficial mistakes made when manufacturing tools; and 3) controlling fire. This bargain was promulgated on a relationship requiring the energy used by a warm-blooded creature doing nothing (the basal metabolic rate, or BMR) to be proportional to the animal's surface area (the amount of skin). This makes perfect sense, for even when a creature is apparently doing nothing, its heart is beating, its neurons are firing, and its digestive system is busily breaking down food into valuable nutrients to be absorbed by the body. These organs are using energy and—by the Second Law of Thermodynamics—generating heat that must be removed if the creature is to avoid overheating. The simplest way to get rid of this heat is to allow it to dissipate heat passively through the skin. Of course, an active system to remove heat could also be used, such as our system of sweat glands, but imagine how inefficient this would be. We would be using more energy to cool ourselves, and we would always be sweating profusely, even while sleeping, and consequently would need more food and would have to drink gallons and gallons of water every day. Accordingly, nature pins BMR to surface area, utilizing a backup cooling system when animals generate more heat than can be removed by ostensibly passive means.

The total energy used by the heat-generating organs, including the brain, cannot increase unless early *Homo erectus* evolves more skin. You may jump to the conclusion that this is why *Homo erectus* was so tall. But no, getting bigger also entails bigger organs, which consume more energy and decrease the proportional energy available for the brain. The only way to get more energy for a bigger brain, if the total energy cannot increase, is to take it from another energy-hungry organ—the heart, lungs, or digestive system. For a creature that depends on endurance running for its livelihood, shrinking the heart or lungs is a nonstarter. This left the digestive system, which became smaller and required less energy because the brain was providing more digestible, energy-rich animal protein and fats. The digestive system and brain had come to an understanding. The energy normally allotted to the gut was reallocated to the brain; in return, the work required to digest food was made a little easier. This supplied enough energy to support a 1,000 cc brain—one not yet big enough to be imaginative. Progress toward a "what if" mind had seemingly come to a stop. Fortunately, approximately 1.3 million years into its evolution, the brain of *Homo erectus* learned to cook.

I've already mentioned findings indicative that *Homo erectus* constructed campfires 1.6 million years ago. However, the evidence for cooking over fire pits begins only 400,000 years ago, at the French site of Terra Amata. Here were found the remains of an early barbecue—charred bones and a simple hearth. Less ancient artifacts from our culinary past are found all across Europe, and by 100,000 years ago, abundant evidence exists of regular fire use and cooking by archaic *Homo sapiens*.

What is significant about the 400,000-year-old find at Terra Amata is its proximity to the emergence of *Homo sapiens,* with a brain as much as 40 percent larger than that of *Homo erectus.* Supporting this encephalization required additional energy that, once again, could only be supplied by the gut—and then only if the energy needed for digestion could be reduced. This is exactly what cooking accomplished: Reducing the metabolic demands of the digestive system freed up energy to expand the brain, and over many generations, *Homo erectus* became *Homo sapiens.*

At this point, the drama becomes muddled in a contentious debate concerning exactly how early *Homo sapiens* evolved to become anatomically modern humans: *Homo sapiens sapiens.* Two extreme scenarios have been proposed. The first is the "multiregional model," positing that all modern humans evolved in parallel from local populations; for example, modern humans in Asia evolved from Asian archaic *Homo sapiens,* and modern Africans evolved from an earlier African population. Some intermixing between the regional populations assured the development of a single species while retaining regional characteristics. The second, or "Out of Africa" model, proposes that all modern humans descended from an isolated population of early humans that evolved in Africa about 200,000 years ago. These early humans then migrated, replacing local populations of archaic *Homo sapiens* as they spread about Africa and Eurasia. Regardless of how anatomically modern humans came to be, fossil evidence documents their presence in Africa 130,000 years ago and in the Near East 90,000 years ago.

Most curious, however, is that very little changed following the emergence of modern humans. Stone tools retained the same basic design characteristics as those made by archaic *Homo erectus.* That is, until 40,000 years ago, during the Upper Paleolithic period, when an explosion of new tools differing in form, materials, and complexity emerged. In Africa as well as Europe, sewing needles, barbed projectile points, and fishhooks carved from bone, antler, and ivory appear. Whereas in preceding millennia, tools and implements showed only global commonality, now each bore a unique regional signature. New products sprang into existence all across Europe: rope, meat-drying racks, stone containers, temperature-controlled hearths, and complex dwelling structures. At the same time, fully mature representational art appears, though its arrival is not presaged by an earlier adolescence. Beautifully carved images of animals and women have been found in Upper Paleolithic sites. Thirty-two thousand years ago, at the German site of Vogelherd, a craftsman carved a small horse in ivory, realistic from its flared nostrils to its curved haunches and swollen belly.

As James Shreeve put it in *The Neandertal Enigma: Solving the Mystery of Modern Human Origins:*

> It is hard to overestimate the significance of this transformation. By all appearances, the people of the Upper Paleolithic came into an innocent, unexamined world and galvanized it with symbol, art, metaphor, and story. They did not simply invent better means of surviving. They invented meaning itself.

How do we explain this flurry of creativity 60,000 years after the emergence of modern humans? Again, the rationales are legion. Some believe that a biological event, unassociated with an increase in brain size or structure, may be responsible. Others suggest that this flourishing coincided with the development of speech. No matter the cause, 40,000 years ago our ancestors developed an imagination. They learned to ask, *What if?* And at the same time, our dependence on energy took form.

Our Energy Past

11

THE GREAT ENERGY RULE

First Came Agriculture

From Big Bang to atoms and heat, from atoms to molecules and heat, from molecules to life and heat, from life to fossil fuels and heat, energy had traveled through both time and distance and was ready to take one more step in its journey of change. Some of it would become the base upon which all other brainwork is built: knowledge. With this base, we have transformed ourselves from the imaginative but humble creatures of 40,000 years ago into more sophisticated, energy-dependent human beings. Yet learning and discovering did not come all at once. It takes time: time to imagine, time to consider, time to build, and time to communicate ideas and thoughts. And we haven't always had time to spend in these activities.

Thousands of years ago, humankind was devoted to surviving rather than questioning. The knowledge we accumulated then was directly relevant to staying alive: When and where do the berries ripen? Today the situation is reversed (at least in the developed world). We spend more time engaged in activities that, at best, are only tangentially related to our survival. We may exercise discretion as to how we use our time. In the past, we had no such discretion, but now we do. How is this transformation to be explained? The answer lies in the laws of thermodynamics.

Life is sustained only as long as the energy derived from food exceeds the energy the organism used to obtain it. This Great Energy Rule, as I call it, is straight from the First Law of Thermodynamics. A cheetah has only the energy available from its last meal or two to expend in chasing down its next, and antelopes use energy from grass and leaves to carry on grazing. Even plants must take in more energy from light than they expend growing leaves to catch the sun's rays. If this was not true, then energy would be created—which is a violation of the First Law. Any life-form that challenges the Great Energy Rule by too narrow a margin and consumes too little energy will grow sick and die.

The ruthless enforcement of the Great Energy Rule demands discretionary hours come from reducing the time spent finding and catching the food needed for survival. For example, humans require approximately 2,000 calories a day to stay alive. Consuming this energy in a half hour by picking up a super-size #3 combo at McDonald's leaves 23.5 hours to use as we see fit (at least insofar as the laws of nature are concerned). On the other hand, harvesting the grain, grinding the wheat, baking the bread, preparing a secret sauce, and slicing the potatoes for french fries would leave far less time to enjoy the day. In addition, the energy expended with all of that slicing and dicing would now have to be compensated through additional consumption.

Our ancestors took a dual approach to increasing discretionary hours. First, they found time-saving techniques for securing food. This common approach is the one most animals use: the giraffe's neck and the elephant's trunk are evolutionary adaptations for more efficient food gathering. Second, our ancestors found an exception to the Great Energy Rule by doing something remarkable: They discovered that more energy *could* be expended obtaining food than it provided—as long as it wasn't the food gatherer's energy. Exploiting this "loophole" is what started humankind on the path to energy dependency.

The first indication that the Great Energy Rule was slightly flawed came when we began to cook. The fire could potentially consume more energy than the cooked food would yield, but as long as the yield was more than the energy expended gathering wood, starting a fire, and preparing the food, the cook was in no danger of dying. It would be thousands of years before another exception was discovered, but in the interim we became more efficient food providers.

The initial step to more effective gathering was the formation of social groups. Infants, the elderly, sick, infirm, and those with special talents were individually exempt from the Great Energy Rule, so long as others expended the energy necessary to hunt and gather. Now the group as a *whole* had to harvest more energy than its members collectively expended. Next came changes that decreased the work of subsistence. Many of these changes were small—better tools for hunting, gathering, and cooking—but one change in particular set us down a road from which there was no return: agriculture.

Around 8000 B.C. in the Fertile Crescent (the area of the Middle East watered by the Nile, Jordan, Euphrates, and Tigris Rivers), humankind grew and harvested crops. The transition from hunter-gatherers to farmers was remarkably fast, requiring no more than a few centuries. What made this conversion so swift may have been the variety of wild flora and fauna native to this geographically restricted area. Here grew the wild progenitors of the founding crops of agriculture: emmer and einkorn wheat, barley, pea, lentil, bitter vetch, chickpea, and flax. Ancestors to the first domesticated sheep, pigs, goats, and cattle also grazed here. From the richness and abundance of the Fertile Crescent, an agricultural package capable of meeting all of humanity's basic material needs developed and spread across Europe. As it spread, native plants and animals were assimilated to yield new crops more suited to local climates. Wild *Aegilops squarrosa,* a grass growing on the shores of the Caspian Sea, crossed with cultivated emmer to produce bread wheat, the most valuable single crop in the modern world. In only 4,000 years, agriculture had spread from the Middle East as far north as the southern end of Scandinavia.

The energy returns from early agriculture were not appreciably higher than those from foraging. In some cases, farming actually required greater energy investments than those of concurrent hunter-gatherer cultures. Crop cultivation supported larger population densities, however, and provided a more reliable source of food. At the same time, not burdened with following the seasons in pursuit of migrating animals and fruiting plants, it became easier to have large families, accumulate possessions, and live in permanent settlements.

Because the Great Energy Rule (sans the energy in the cooking process) was applicable to early agricultural societies, the work of food producers provided the energy to sustain the nonfood producers who administered, defended, and built their settlements. Thus, agricultural efficiency determined how many soldiers, bureaucrats, and craftspeople a society could support. Flourishing agriculture propagated through cultures and facilitated prosperity in all facets. It's no wonder that agricultural productivity became prime fodder for those who now had the discretionary time to imagine.

Though advances in planting, harvesting, and storage of all crops were significant, it was in grain cultivation where the greatest rewards would be realized. Grain is the dominant agricultural crop, worldwide. While foraging societies relied on tubers, seeds, and fruits to provide most of their food energy, these options were limited in places where cultivation was extensive and plant products were stored for later consumption. The water content of fruits and tubers makes them susceptible to fungi, bacteria, and insects, which means they're difficult to preserve. Even when that is overcome, their tremendous bulk poses an equally formidable storage challenge. Seeds, with lower water content, can be more easily stored; of these, it is cereals—wheat, rice, and corn—that have the greatest energy yields.

In terms of nutrition, cereals are remarkably similar, with the bulk of their energy coming from easily digestible starches (carbohydrates). The remainder is in the form of protein, which accounts for between 10 (rice) and 15 percent (wheat) of its energy content. However, the primarly role of protein in human nutrition is not as an energy supplier, but rather as a source of the amino acids needed to build and repair body tissues. Cereal grains do not provide all of the amino acids required by human beings, though animal products—milk and meat—do. Cereals fall short on lysine, and legumes are deficient in methionine. Consequently, vegetarian societies stumbled upon a solution to this nutritional puzzle by consuming a combination of seeds. In China, soybeans, peas, and peanuts supplemented wheat and rice; in Europe, the combination was peas and beans along with wheat, barley, oats, and rye; and in Mesoamerica, beans and corn.

The first energy-intensive step in the cultivation of cereals is planting, which begins with field preparation. Originally, this process involved hoeing or the use of a scratch plow, which was nothing more than a pointed stick with a handle that opened a shallow furrow for seeds. Later, these plows were tipped with metal, reducing the energy required to pull or push the plow through soil. But even then only the lightest soils could be tilled, as humans lacked the power to push plows through even moderately heavy loams, a task that is more suitable for many four-legged animals.

Middle Eastern donkeys were likely the first draft animals, plowing the fields of Mesopotamia around 3500 BCE. Donkeys, being small, could generate only a bit more sustained power than a large man. In general, plowing force is proportional to an animal's body weight. Hence, by 3000 BCE, oxen were the draft animals of choice. Horses are heavier yet, and more powerful, but they were not used for plowing until the first century BCE. Harness and yoke designs slowed their introduction to the agricultural workforce by 1) not taking full advantage of the larger animal's power; and 2) causing the horse to choke. The Chinese collar harness eliminated these problems.

With a collar harness and horseshoes, a horse could plow 30 to 50 percent faster than an ox (except in heavy, wet soils). This advantage accounted for the spread of horses as draft animals, and by the twelfth century CE, European agricultural production was dependent on horses.

The use of draft animals does not constitute another exception to the Great Energy Rule, however. Donkeys, oxen, and horses must be fed. A horse's hay and coarse grains require cultivated land for their production. Oxen, not as picky as horses, are able to subsist on roughage and straw, though their energy output cannot exceed their intake. What draft animals offered was a time-saving measure. With a hoe, 100 hours of work are required to ready one hectare (2.471 acres) of land for planting grain. But with an ox pulling a simple wooden plow, the same task can be completed in 30 hours. In essence, the draft animal's "discretionary time" was diverted to our use.

Our expansion into nonfood-producing activities was only possible through our escalating dependence on animal power. In the fourth century, animals provided 90 percent of the energy used to provide food for hungry Romans. By the beginning of the nineteenth century, draft animals supplied 96 percent of the work needed to plant and harvest the crops of western Europe. A hundred years later, the human labor involved in planting and harvesting was essentially negligible. Brainwork had replaced muscle work as the farmer's chief activity.

In the absence of "animal unions," the growing draft-horse workforce was responsible for spectacular gains in energy efficiency. In the Roman and early medieval years, forty times as much wheat energy was reaped from the harvest than the energy invested in its production (feeding the animals and the farmer). In western Europe at the beginning of the nineteenth century, the ratio was 200:1, and by the end of the same century, the ratio was commonly above 500, and in some instances above 2,500. When seed for the next year's planting and storage losses were included, net energy gains ranged from twenty-five times for medieval to two hundred times for early-twentieth-century farms.

Although draft animals more than paid for themselves with increased agricultural productivity, they also came with a significant cost: land. In the first two decades of the twentieth century, when the country's human population was approximately 75 million, its horse and mule population was close to 25 million. Of these, about 70 percent were part of the agricultural workforce, living on the farms and ranches of the United States. Fully one-quarter of America's cultivated land was devoted to feeding our equine partners.

Human populations were linked to those of draft animals. Though conceivably agricultural output could be increased, it was only possible through the work of more draft animals, which required that more land be turned over to their maintenance. The only way to escape these constraints was to find a power source in place of draft animals. To some extent, inroads in this direction had been made. Wind and water were being exploited to power grain mills and drive irrigation systems, instead of using animals to do the same work. But, as a fraction of the total energy employed for agricultural production, these were insignificant contributions. The Great Energy Rule had kept humankind in check: population was checked, the ratio of nonfood producers to food producers was checked, the number of people who could be pulled off farms and sent to war was checked, and most important, the time a society could spend doing brainwork was checked.

To break loose from the tyranny of the Great Energy Rule, we had to boost horsepower without resorting to more hungry horses. And we had just the thing: beasts of burden that didn't eat grain grown on precious agricultural land but that drank oil pumped from beneath it. Tractors, combines, threshers, and trucks moved into the barns where Clydesdales, Percherons, Belgians, and mules had once lived and worked. As far as humankind was concerned, internal combustion engines had reduced the Great Energy Rule to something that applied to *other* living things—not us.

The transition from horses to horsepower was as significant as the switch from foraging to farming. Farmers did not instigate this transition, nor did governments looking to feed a growing population encourage it. Rather, it was craftspeople with time to imagine and time to wonder because efficient farming had freed them of the fundamental need to forage and feed themselves.

12

A TALE OF TWO CITIES

Wood to Burn

This tale of two cities, like that of Dickens's, is set in an age of wisdom and foolishness. Agriculture brought the freedom to imagine and learn to the people of our two cities, which were separated by culture, climate, oceans, and time. Imagine and learn they did. But they were also imprudent, because they did not recognize the limits that energy placed on their imaginations. In one case, these limits contributed to the collapse of an entire culture. In the other, a great city was rescued from oblivion by pure luck.

Tikal was a major cultural and population center of the Maya, located on the Yucatán Peninsula. What is left of the city that once covered 60 square kilometers (23 square miles) is hidden beneath the canopy of the Guatemalan forest. Huge open stadiums where audiences thrilled to the exploits of young men playing an ancient ball game now sit empty and moss-covered. Hand-carved stone monuments are pitted and weather-worn. Reservoirs, holding enough water to meet the needs of ten thousand people for eighteen months, are overgrown with native plants. The city's only permanent residents are the howler monkeys, colorful birds, jaguars, and other wild animals that seek shelter among its ruins.

Twelve hundred years ago, Tikal was home to 100,000 inhabitants. Maize fields, cleared of native vegetation presumably through slash-and-burn techniques, stretched to the horizon. Dotting the cornfields were smaller plots of green beans, chilies, pumpkins, tomatoes, and avocados. By European and Asian standards of the time, Mayan agriculture was primitive. Without draft animals, plows, or carts, farmers used planting sticks to make holes in the soil for seeds. Still, these agricultural techniques were sufficient to support significant ruling, religious, craftsman, and intellectual classes, which constituted perhaps 30 percent of the population.

The elite classes helped to build a vibrant culture with a written language and a numbering system that included the concept of zero. At the top of the ruling class was a king, who also functioned as high priest. It was his responsibility to keep the Mayan people prosperous by bringing rain and bountiful harvests—a job made easier by virtue of family ties to the gods and his own personal divinity. The king also presided over rituals associated with astronomical events; astronomers, who tracked the sun through a 365-day solar year and kept accurate tables of solar eclipses and the positions of celestial objects, including Venus, the war star, predicted the timing of these events.

Mayan kings were constantly at war with their neighbors. The king of the losing side was often taken captive and tortured, sometimes for years, before being dispatched. Thanks to the Mayan propensity to enshrine their barbarous acts in monuments and murals, we know something of Mayan torture. Captives were mutilated; sometimes their fingers and lips were cut off, their lower jaws removed. After tiring of a captive's suffering, the Maya did not kill their victims to relieve misery or spare additional humiliation. Instead, they slaughtered with an equally terrible act of violence, such as tying the captive into a ball and then rolling him down the steep stairs of a temple.

A counterpoint to the brutality of Mayan warfare is the glorious architecture for which the Maya are most remembered. Every year, thousands of tourists marvel at the ruins of Tikal where six huge pyramids (60 meters/200 feet high) topped with temples are silhouetted against the tropical sky. Isolated from the violence that occurred here, these edifices evoke a sense of almost sublime accomplishment. There are also immense stone palaces, residences, and monuments. Surprisingly, these magnificent structures were built without the benefit of metal tools or technologies that would have assisted with the movement of large stones, like pulleys or even a wheel.

The Maya apparently made do with an abundance of manpower and the local limestone, which could be worked with stone tools. Limestone served not only as a structural material, it could also be processed to make a cementlike mortar—though that was hardly necessary since the masons of the Classic Maya period (200–850 CE) could cut stones so precisely as to lock together under their own weight. However, unlike other stoneworking cultures, in the Maya's it had become fashionable to finish the pyramids, palaces, and other structures with stucco, which eroded quickly and required frequent repair or replacement. The problem? Producing stucco was energy intensive.

Stucco is made from a mixture of quicklime and water. Quicklime is prepared by heating limestone to roughly 1,000°C (1,800°F). At this temperature, the limestone (calcium carbonate) breaks down into quicklime (CaO) and carbon dioxide.* Plaster and stucco had been manufactured at least since early Roman times. In the Mayan process, the fuel of choice was wood or other organic matter with average water content greater than 50 percent. High-moisture fuel simplifies the design of the limekiln but reduces its efficiency. It has been estimated that, by volume, ten parts of freshly cut wood were needed to produce one part quicklime. The forests surrounding Mayan cities also had to provide the fuel for domestic uses. With these demands, along with the clearing of terrain to provide more agricultural land, the Yucatán Peninsula was being steadily deforested. Accompanying this was erosion and perhaps a man-made drought, since reduced forest transpiration decreased the cycling of water vapor to the atmosphere. Around 869 CE, when the last known Tikal monument was built, the city settled into a two-hundred-year waning. By 1050, the city was abandoned. This prolonged collapse took place in the Classic Mayan period, proceeding at somewhat different rates throughout the empire.

Skeletal remains from various sites indicate that during the collapse, the Maya suffered from health problems associated with malnutrition. When Cortés crossed the Yucatán in 1524, the Mayan population had been reduced from as many as 14 million to as few as 30,000. His conquering army nearly starved, unable to find villages where they could acquire food. None of the great Mayan cities were still inhabited. Cortés passed within a few kilometers of Tikal, but he saw nothing. The jungle from which it was torn had reclaimed the great city.

The Maya must have known that they could not carry on denuding their landscape. Fathers must have told sons about the time when the forests came right up to the door and they spent less time gathering

---

*This is the same reaction we encountered in chapter 8 when discussing the carbon cycle. Limestone is heated by friction during subduction, leaving carbon dioxide dissolved in molten magma, which is returned to the atmosphere via volcanoes.

wood and more time making stucco. In spite of the evidence, they continued to stucco pyramids, cut trees for fuel, and clear agricultural ground. The cataclysm from the resultant deforestation desiccated the land and drove the empire to starvation and oblivion, its survivors scattered to the hills. The question we cannot help but ask is *Why?* Why would a culture proceed down a path that was clearly taking them to a disastrous end? Unfortunately, Mayan records are so sparse that we can only guess what the peasants and nobles thought as their city and civilization declined and, ultimately, vanished. In the the second city in our tale, however, record keeping was far superior, giving a glimpse into the minds of its inhabitants as they repeated the mistakes made by the Maya centuries earlier.

When Tikal was at its height and its 100,000 inhabitants walked among fabulous stucco-covered stone buildings and monuments, the city of London was a prosperous trading center that had been continuously inhabited since 43 CE. It was then that invading Roman troops marched on the economic center of Britain, Colchester. They were halted by the Thames, but it was a temporary annoyance, since the troops built a bridge, which formed the hub of the new network of Roman roads they soon made to link the country. The settlement on the north riverbank near the bridge, called Londinium, grew and prospered, reaching a peak population of about 45,000. But the Roman Empire was not eternal. Toward the end of its preeminence, it could no longer defend London's trade routes, and at the end of the fourth century the city began a several-century decline.

But London was too well situated as a center for commerce to be thwarted for long. By the seventh century, trade had once again expanded and thrived for nearly two hundred years. But with prosperity came unwanted attention. In 851, with a force of over three hundred longboats, Danish Vikings sailed up the Thames, sacked the surrounding countryside, and burned the city of London to the ground.

After it was rebuilt, the city changed hands regularly. Passed forcibly back and forth between the Anglo-Saxons, Danes, and Normans, London became the largest city (at about 18,000 inhabitants) and the most prosperous in England by 1066, when William the Conqueror (a Norman) was crowned King of England in a ceremony at the just-completed Westminster Abbey.

The industries that would transform London into one of the world's great industrial centers were just emerging as William ascended to England's throne. However, the craft that generated the energy for these nascent industries was well ensconced. As with the Maya, this energy was afforded by trees, but in England, tree wood was often converted into the hotter-burning charcoal, a practice that dated back thousands of years.

Today, with the proliferation of the (grossly inferior) gas grill, fewer and fewer people use charcoal for barbecuing. Given the minor role of briquettes (which have little in common with the real stuff), it might be difficult to believe that charcoal was, until relatively recently, one of humankind's principal energy sources.

As part of my seventh-grade science class, we made charcoal. The activity involves heating scraps of wood in a test tube with a Bunsen burner or other open flame. The narrow opening of the test tube (which can be made narrower with a glass tube in a one-holed rubber stopper) prevents air circulation; the wood is deprived of oxygen and cannot burn. Instead, it gets hot, and as the temperature increases, the volatile wood components—composed of water and various liquid organics, like turpentine and tars—are driven off and can be collected with additional apparatus. Over time, what is left behind in the test tube turns black, eventually becoming almost pure carbon, that is, charcoal.

Test tubes, stoppers, and Bunsen burners were of course not available to medieval charcoal burners. Instead they employed a technique that dates back to ancient times, in which split wood was stacked on end to form a tightly packed conical pile as large as 10 meters across and three meters high. Air openings were left at the bottom, along with a central shaft serving as a flue. The whole structure was then covered with turf or moistened clay. The wood around the bottom of the flue was ignited, spreading the combustion out and up through the pile and leaving behind charcoal from the oxygen-starved combustion.

The efficiency of charcoal production depended sensitively on the size of the charcoal pit, which controls the rate of combustion. On average, 100 parts of wood yielded about 60 parts by volume, or 25 parts by weight, of charcoal. Smaller pits were only about 50 percent efficient (by volume, 50 parts per hundred wood), though large pits, where combustion was slow and severely oxygen deprived, could be as much as 90 percent efficient.

The production of charcoal was a craft, requiring great skill and knowledge. Medieval charcoal burners often worked in solitary groups and were known for their questionable business practices: A sack of charcoal was priced at about a day's wages and it was common to find it adulterated with rubbish.

Wood and charcoal were the oil and gas of the Middle Ages, used as fuel for cooking and heating, to produce salt from seawater, to dry hops and other grains, and to manufacture lime, glass, and steel. The production of steel, in particular, demanded large supplies of charcoal.

In medieval Europe, steel was produced in a bloomery, a furnace in which iron ore is heated with burning charcoal. Oxygen was supplied to the nearly airtight furnace via bellows, which elevated the temperature to nearly 1,300°C (2,300°F); carbon monoxide from the burning charcoal then reacted with the iron ore to make metallic iron. But because the furnace was not hot enough to melt the metal, it collected at the bottom in a spongy mass called a bloom. Slag, ash, gases, and other impurities were trapped inside the bloom and were removed by reheating the iron to about 1,100°C, again in a charcoal-fired furnace. At this temperature, the slag was soft and could be "wrung" from the iron by a laborious process of folding and beating. This process was repeated multiple times—folding, heating, and beating—until most of the slag was squeezed from the bloom, leaving behind the much softer and more malleable "wrought iron." The next step, turning soft wrought iron into hard steel, was also energy intensive. Generally, the soft metal is fashioned into its desired shape and then heated with charcoal dust in the absence of oxygen. Carbon atoms diffuse into the iron surface to make steel.

The medieval steel-making center of England was located just south of London in an area known as the Weald. Hilly countryside unsuitable for large-scale farming, the Weald was covered with great hardwood forests of oak, beech, and chestnut. A belt of iron-bearing ore wound through the terrain, randomly thrusting through the surface. Streams were dammed for power to drive hammers and bellows.

The demand for iron and steel in medieval England was modest. Occasionally the king would place large orders and the ironworks of the Weald would temporarily shift into high gear. In 1242, King Henry III asked the Archbishop of Canterbury to provide, from his estates in the Weald, 8,000 horseshoes and 20,000 nails. The largest-known order came in 1253 when the Sheriff of Sussex supplied the king with 30,000 horseshoes and 60,000 nails. The forest of the Weald was the chief provider of charcoal supporting England's iron industry.

This changed with the expanding use of cannons, which by the early fifteenth century had rendered castle walls as protective as tissue paper. Improved gunstones—what we call cannonballs—contributed to the increasingly destructive power of cannons. Originally carved from stone, cannonballs evolved into spheres of cast iron, a technology made possible by furnaces that could melt iron. The molten metal from these blast furnaces was then cast to shape in a mold. The first blast furnace in England was constructed about 1491 in the Weald and was soon employed in the production of gunstones.

It was none too soon, for in 1509 at the age of seventeen King Henry VIII came to the throne of England and almost at once began picking fights with his neighbors Scotland and France. In 1512, England declared war on France. The armaments necessary to support the impending war were to be made in the Weald, where in 1513 a single ironworks supplied King Henry's army with nine tons of iron gunstones.

From casting cannonballs to casting cannons, technology marched. By the mid-sixteenth century, cast-iron cannons of the Weald were the finest in the world. To satisfy the demand, fifty-three forges and blast furnaces operated there in 1549. Though the iron produced was not all for military use, it did represent a significant fraction and was a driving force for expansion. Twenty-five years later the number of furnaces and forges had jumped to 110, producing several thousand tons of iron, including several hundred tons of cannons. While the Wealden forest had been ample to support early-medieval iron manufacture, it was strained to the breaking point under the expanding call for iron and other products reliant on wood. Principal among these products were ships, particularly those of the expanding English navy.

Built from oak, the large warships of the early sixteenth century consumed the timber from as many as two thousand hundred-year-old trees. The most attractive trees were located near rivers or canals, where water transport lessened the cost of moving great quantities of wood. And guess what? The best place to find such oak trees was south of London, in the Wealden forest.

England was impaled on the horns of a dilemma. Great oak trees could be reserved for shipbuilding, depriving the iron and other manufacturing industries of an energy resource. Or cannon manufacture could continue, possibly consuming the trees needed to build and maintain a navy. In 1544, Parliament tried to have it both ways, passing a law mandating that loggers leave a certain number of mature oaks, the latter-day counterpart of a "strategic reserve." In addition, forests were not to be converted to agricultural use. Due to its location and economic importance, however, the Weald was excluded from the law's provisions—an exemption that proved not popular with many of those who lived and worked near its ironworks.

In 1547, the Duke of Somerset, in his capacity as Protector of the Realm, appointed a commission to look into the question of wood, timber, and the iron industry. Public hearings were held where witnesses expressed their concerns and the commission recorded their words. According to an inhabitant who lived near fifty ironworks in the south coast port of Hastings, each one used more than 1,500 loads of wood per year. The uncontrolled harvesting of this wood was causing collateral damage to the forests and increasing the cost of wood, which had doubled in just fifteen years. By some accounts, the forests were being depleted so fast that soon there would be no wood for houses, bridges, boats, gunstocks, arrows, wheels, or barrels.

Richard Cowen writes that one petitioner implored the commission to close the ironworks, or else the ports of Hastings and Rye would

> lack necessary wood for fuel for the relieving of poor fishers after their arrival from their daily fishing to dry their clothes and warm their bodies, by whose trade chiefly the said towns stand, the same will shortly decay.

Like many of today's congressional hearings, the citizenry vented their rage and frustration but no action was taken in the end. Perhaps the economic and political might of the ironmasters had allowed them to head off anything as drastic as closures. The locals expressed their anger over the inaction in a brief episode of rioting and sabotage, which played into the hands of the Duke of Somerset's enemies. He was removed as Protector in 1552 and charged with treason for, among other things, instigating "sedition, insurrection, and rebellion." The duke was executed the same year.

England's precarious relationship with its natural resources grew increasingly uncontrollable with the mounting militarism of Spain under its devoutly Catholic king, Philip. He envisioned a Catholic Europe and, in 1572, to repress a rebellion of Protestants in Spain's Flemish provinces, he deployed 67,000 battle-hardened soldiers to Holland. By 1574, the number had grown to 87,000. At a time when Protestant England could raise an army of less than 30,000, England's survival depended on its ability to stop an invader in the Channel, a charge that required English ships and cannons—both provided by the forest of the Weald.

Finally, all hope of a Catholic England died with the 1587 execution of Mary, Queen of Scots, and Philip instantly readied an invasion armada. Due to a combination of factors, including the weather and more maneuverable English ships and better cannons, the Spanish Armada was destroyed. England became the new master of the seas, initiating an era characterized by even more shipbuilding, cannon making, and deforestation.

To be fair, the iron and shipbuilding industries were not fully responsible for wood shortages. The building trades consumed timber, as did brewing, glass manufacture, lime production, and salt making. There were over three hundred wood-burning furnaces in the salt-producing town of Cheshire. One London brewer grumbled in 1578 that his trade, which at the time was consuming 20,000 wagonloads of firewood a year, could not continue without more and cheaper wood.

In Elizabethan England, wood and charcoal prices grew faster than those of any other major commodity. This inflation hit the most populated regions the hardest; in London and the surrounding area, wood and charcoal were driven beyond the means of the average worker. These soaring costs began to transform London from a city built of wood to one of bricks, which provided only partial relief: Bricks were fired in charcoal- or wood-fueled furnaces.

By the beginning of the eighteenth century, wood had become the resource driving the cost of everything else. Charcoal represented

80 percent of the costs associated with iron production. England, having harvested its own forests, was increasingly dependent on foreign sources to provide its "wood fix." Ireland had become a barren island by 1650, its forests of oak and other trees felled to supply the insatiable needs of England. The first product exported to England from the American colony of Jamestown was timber. The Massachusetts state capitol building is topped with a gilded pinecone to symbolize the importance of the lumber industry to an early New England economy that exported much of its timber to England. But imports could serve only as a stopgap—deforestation was sweeping Europe. England was simply among the first countries to feel its effects.

London in the eighteenth century was riding the same downward spiral as Tikal. Its sole source of energy was being rapidly depleted to give substance to an expanding set of products and ideas born within our imaginations. Today, walking among the remains of the Wealden iron industry, ruined forges, furnaces, slag heaps, and ponds that supplied power for bellows and hammers are scattered about the countryside. From these ruins, one might conjecture that a once-great city like Tikal lay just over the hill. But cresting this hill, the view is not of a dead city but of London, still one of the world's great industrial centers. What saved London from the fate visited upon Tikal was an accident of geography.

13

SAVING THE FORESTS

Coal to Burn

Coal was used sparingly for thousands of years. There is some evidence it was employed in ancient China to smelt bronze. The Romans used it for heating, and in the 1300s, the Hopi Indians of North America used coal for cooking, heating, and to fire pottery.

Compared to charcoal, coal is a nasty fuel. It produces thick black smoke and smelly fumes. On the positive side, it tends to "cake" into a lump that does not blow about when fanned with a bellows, as do the ashes and small embers in a charcoal fire. As far as smithies were concerned, this was an advantage; they could be less concerned about coal spreading fire, a benefit that outweighed its noxious fumes (situating the forge downwind from the village mitigated the stench). The

English coalfields of the thirteenth and fourteenth centuries were mined only to the extent necessary to supply blacksmiths; coal had been ignored as an energy source except where wood was in short supply. But wood was now growing scarcer, and the situation was growing desperate in the large manufacturing centers, which made London ground zero of the wood shortage and the heartbeat of the coming "alternative energy" revolution.

Along the banks of the Tyne River in the northeast of England, seams of coal erupt through the soil. In the early part of the thirteenth century, this "black earth similar to charcoal" was mined and transported to London on ships departing the port city of Newcastle. There, it was sold as "sea-coal." By 1300, several thousand tons of coal shipped to London via Newcastle each year. The growing trade was fed by the escalating cost of charcoal. London limekilns were switching to "cheap" coal as early as 1280, prompting complaints from the city dwellers who had to breathe the polluting by-products. In parts of London, the air grew so foul that a royal proclamation prohibited coal burning by the south London lime industry:

> An intolerable smell diffuses itself throughout the neighboring places, and the air is greatly infected, to the annoyance of the magnates, citizens, and others there dwelling and to the injury of their bodily health.

No relief came, as profits continued to rank over environmental concerns, prompting authorities to pursue more persuasive policies. Offenders were threatened with "great fines and ransoms" for first transgressions, and with the destruction of their kiln for a second. Even so, London's air grew progressively rank.

As the cost of charcoal surged, one industry after another shifted from charcoal to coal. By the mid-sixteenth century, London's air was so burdened under a cloud of coal smoke that Queen Elizabeth I banned coal burning while Parliament was in session. Richard Cowen writes that the queen was so "greatly greaved and annoyed with the taste and smoke of the sea-cooles" and troubled by the "noysomme smells" that brewers near her London palace offered to burn wood in place of coal.

With even royal displeasure having little effect on coal's proliferation, the government gave up trying to regulate its use. In 1661, John Evelyn wrote an essay titled "Fumifigium" about the

> Columns and Clouds of Smoake, which are belched forth from the sooty Throates . . . rendering [London] like the approaches of Mount-Hecla. That hellish and dismal cloud of sea coal [means] that the inhabitants breathe nothing but an impure and thick mist, accompanied by a fuliginous and filthy vapour, which renders them obnoxious to a thousand inconveniences, corrupting the lungs and disordering the entire habit of their bodies, so that catarrhs, phthisics, coughs and consumption rage more in that one City than the whole Earth besides. . . . Is there under Heaven such coughing and snuffling to be heard as in the London churches where the barking and spitting is incessant and importunate?

Paradoxically, in some parts of England the expanding demand for coal contributed to deforestation. Trees were harvested to support a coal industry that needed timber to shore up mine walls, as well as lumber for the docks, wharves, barges, and merchant vessels that were part and parcel of London's coal trade. The loss of forests prompted this response from William Harrison in the 1580s:

> Of coal mines we have such plenties as may suffice for all the realm of England. And so they must do hereafter indeed, if wood be not better cherished than it is at present.

The challenge to the English authorities of the late sixteenth and early seventeenth centuries was to maintain forests for their vital importance while recognizing that burning coal was the only means to achieve this end. By 1615, England's security was tied to a policy encouraging replacement of fuel wood with the only available alternative. Henceforth, brick chimneys sprouted like weeds from London rooftops. These spewed forth sooty smoke, fouling not only the air but the chimneys. In 1618, it took two hundred chimney sweeps to keep London's fireplaces clean and functioning. Constant contact with the chemical-laden soot made the sweeps susceptible to cancers of the lungs and skin.

Coal's carcinogenic soot results from the copious impurities it contains, including sulfur and mercury. Sulfur is particularly prevalent in coal, and when burned it combines with oxygen to form sulfur dioxide. This gas rises into the air, combines with atmospheric moisture, and falls as sulfuric acid, or acid rain. This is not the only by-product of sulfur. As a class, sulfur-bearing molecules are disgusting, primarily because of their odor. Remember hydrogen sulfide smelling like rotten eggs? Garlic and onions owe their well-earned odiferous reputation to sulfur. And skunk spray bears several sulfur-containing molecules known as thiols.

Brewers, principally consumers of wood in their malt-drying furnaces, stayed away from coal since smelly thiols and other equally rancid molecules would permeate the malt, making for a less than tasty brew. Still, coal's comparative cost made it attractive as an energy substitute—if the sulfur problem could be solved. This led Sir Henry Platt to suggest, in 1603, that coal could be processed just like charcoal. In other words, cook the coal in an oxygen-deprived atmosphere to drive off its volatile components. After decades of trying, the brewers of Derbyshire managed to "char" coal to produce a carbon-rich product called coke. Coke is harder, produces little smoke, and burns hotter than coal. More important, as far as the brewers were concerned, coke-dried malt produced great beer.

Sulfur made coal useless not only to brewers but also to iron smelters and steel makers. In this case, it was not sulfur's odor but the sulfur atoms that were the culprit. It takes only a small quantity of sulfur in iron and steel to make them brittle—not a desirable trait in cannons, plate armor, or most iron and steel products. A brittle cannon has the worrying habit of blowing up when used, and plate armor breaks under the impact of a lance. Since brittle iron is not very useful, blast furnaces were fueled with charcoal instead of coal, despite the costs.

It's surprising that even as the English iron industry was suffocating under charcoal's escalating price, no one thought to try coke. That is, until the beginning of the eighteenth century. Abraham Darby had apprenticed in a brass works that made mills for grinding malt; he had visited the breweries and had seen the coke-fueled malting ovens. When he launched his own business, he chose to make cooking pots from cast iron rather than the traditional, and more expensive, brass. He built his pot factory around an abandoned blast furnace at Coalbrookdale in the West Midlands, close to a source of iron ore and coal. He constructed a coking furnace, and on January 4, 1709, the first-ever casting of *iron from a coke-fired blast furnace was poured.* Darby's innovation spread through the Weald's ironworks, and in short order, one of the three critical pillars supporting the Industrial Revolution was in place—coke fueled blast furnaces.

Coal was now a hot commodity—in every sense of the word. Wood, as a fuel, had been replaced. At the same time England, which may have been on the verge of decline, had done an about-face and was on its way to world domination. England was loaded with coal—a refreshing contrast to the paucity of wood (exacerbated by centuries of deforestation and foolish resource management). Blessed now with an abundance of resources, English industries were unconstrained and embarked on a two-century energy splurge. The first stop: Find more energy.

Surface mining of coal was no longer sufficient. To exploit this resource, miners needed to follow coal deep into its subterranean lair, where vast "galleries" of the energy-rich black stone were excavated. The mine shafts were progressively deeper—20, 50, 100 meters—until halted by water seepage. If left unattended, these leaks flooded mines, making further excavation all but impossible.

Dissatisfied with the constrictions of the local water table, pumps powered by muscle—human and animal—were enlisted to keep the mine shafts dry and workable. Then in 1712 Thomas Newcomen invented and installed the first steam-operated pump in a West Midlands coal mine. Using coal as an energy source, water was turned into expanding steam to drive a piston, which lifted mine seepage to the surface. Newcomen's first pump lifted 600 liters (150 gallons) of water up a 50-meter (160 feet) shaft in one minute: an unprecedented achievement. And since the pump ran on coal, it cost the mine owners essentially nothing to operate—which was certainly not the case for muscle-powered pumps. In an instant, the most significant obstacle to cost-effective coal mining was removed; all sources of coal were now exploited; and another pillar of the Industrial Revolution—the steam engine—was operating.

One pillar remained. Coal is heavy and bulky. Mining it was becoming less expensive with the advent of new mining techniques and technological innovations, but getting the coal from the mine to the consumer was another matter entirely. Whereas transport over water was reasonably efficient—at the end of the seventeenth century ships and barges moved more than a million tons of coal hundreds of kilometers—land transport was proving to be prohibitively expensive. Coal was seldom moved more than a few kilometers overland. Hence, coal seams not fortuitously situated along the coast or on inland water-ways went undeveloped.

Bumpy, muddy, and sandy surfaces reduce the amount of coal a horse can pull on a road. That amount increases dramatically if the horse is pulling a cart over tracks, even if those tracks are made of wood. Huntington Beaumont built the first "wagon-way" in 1604. Though he never profited from this innovation, falling into a £30,000 bankruptcy, wagon-ways had a staggering impact on coal production. In 1725, a 30-meter-long and 20-meter-high bridge was built as part of a wagon-way connecting a mine to the Tyne River, five miles away. Seven hundred carts, each laden with four tons of coal, transported their burden to the waiting barges. As the production advantages associated with wagon-ways were felt, mine galleries were fitted with rail systems to facilitate hauling coal to the surface. By 1800, iron rails had replaced wooden ones. Their harder surface reduced friction and made the work of pushing carts much easier—a fact clearly appreciated by the miners, who praised iron rails with the verse:

> God bless the man wi' peace and plenty
> That first invented metal plates
> Draw out his years to five times twenty
> Then slide him through the heavenly gates.

On a wagon-way, a horse can pull four tons of coal. It can drag a 30-ton barge along a river; 50 tons if the water is still, as on a canal. The economics were obvious, prompting English mine owners to embark on fifty years of canal building, beginning in the late 1700s.

One of the most expansive canal systems was that servicing the Duke of Bridgewater's Worsley colliery. The duke commissioned James Brindley to engineer a 16-kilometer canal running directly from his mine to the middle of Manchester. Brindley designed a gravity-flow system that included an aqueduct 183 meters above the Irwell Valley. The canal cut the cost of coal delivery in half. And what worked above ground also worked below: 60 kilometers of underground canals on four levels reduced the cost of transporting coal within the mine.

The third pillar of the Industrial Revolution—transportation— was all too obvious. Steam engines were now pumping water from mines, driving ironworks' bellows and hammers. So why not harness a steam engine to the front of wagon-way carts and canal barges?

Lord Dundas, governor of the Forth and Clyde Canal Company, was the visionary responsible for demonstrating the practicality of steam-driven water transport. In 1803, after several years of trials and improvements, the steam-tug *Charlotte Dundas,* named for His Lordship's daughter and designed by Alexander Hart, towed two 70-ton barges 30 kilometers (almost 20 miles) along the Forth and Clyde Canal to Glasgow. Despite a strong headwind, which stopped all other canal boats, the tug completed the trip in just over nine hours. Twelve years later, steamships were regularly plying the coastal waterways, transporting cargo and passengers between the ports of Liverpool and Glasgow.

Entrepreneur Samuel Homfray seized the opportunity to demonstrate the practicality of steam-driven rail transport. He had the money, saw the potential, and commissioned Richard Trevithick to make it happen. On February 22, 1804, the first locomotive began a two-hour journey, hauling 10 tons of iron, 70 men, and five extra wagons nine miles from the ironworks at Pen-y-Darren in Wales to the bottom of the Abercynnon valley. In the time it took to lay tracks, locomotives were transporting coal, iron ore, and other commodities—as well as passengers—about the English countryside.

Coal was a little slower to catch on in the Americas, where a wood shortage hadn't been felt. The first record of commercial U.S. coal mining, near what is now Richmond, Virginia, was in 1748. Coke did not replace charcoal in blast furnaces until 1875, but the economies of coal eventually made sense, and America, too, converted. No city more typifies this than Pittsburgh.

High-quality bituminous coalfields located near the city provided the energy to make Pittsburgh a leading producer of glass, iron, and textiles. By the mid-nineteenth century, the inhabitants of this growing industrial giant were coping with the same sooty air to which Londoners had grown accustomed centuries earlier, and with similar results. Pleas for government intervention were ignored or ineffective until well into the twentieth century, when other energy sources became available that Pittsburgh would shed its mantle of the "smoky city."

As we saw in London and Tikal, people and industries will not willingly abandon an energy resource without either 1) dwindling supplies; or 2) the presentation of a more efficient, less expensive or versatile alternative. England, much of Europe, and the United States sat on tremendous coal deposits, so supplies were not about to fail. Which means that someone had to discover a more efficient and versatile energy source, one that could be used to power internal combustion engines.

14

Oil to Burn

We seek out light. Perhaps we do this because we are creatures shaped by the Sun; supreme predators of the day, helpless prey at night. Whatever the reason, turning night into day is as much a human activity as is imagining.

The first man-made light source were the campfires and flaming brands that *Homo erectus* used to frighten predators. For a million years, burning wood remained the only source of artificial light. Then, about 5,000 years ago, someone placed twisted fibers into a clay jar filled with olive oil and the lamp was born. This "high-tech" invention changed the ancient world. Lamps were portable, provided hours of illumination, and could be used indoors without generating excessive smoke. The best lamps produced several times as much light as candles, which have been found among the detritus of first- and second-century Roman archaeological sites.

Too dim to sweep even nearby corners clean of shadows, lamps provided only partial relief from the night. Until the late eighteenth century, that is. In 1783, Swiss physicist Aimee Argand invented a new wick. Cotton or other fibers were woven into a tube, allowing more air circulation for the burning fuel, thereby emitting more light.

Hand-in-hand with a better wick came the search for improved fuel. Lamps had used a wide variety of vegetable and animal oils over the years, but whale oil was the illuminant of choice. It burned cleaner and with very little odor. Of all whale oils, spermaceti (from the nose of sperm whales) was most prized. It was also expensive. In the early 1800s, spermaceti sold for roughly $50 per gallon by today's standards. To supply this valuable commodity a thriving whaling industry developed, providing not only oil for lighting but for machine lubricants made from whale blubber as well. Ports along the northeastern coast of the United States became the center of a huge whaling industry, totaling 735 ships in 1846. From ports like Nantucket, New Bedford, Essex, and Falmouth, ships set sail in hunt of whales. At its height in 1856 the New England whaling industry supplied the worldwide market with 4–5 million gallons of spermaceti and 6–10 million gallons of machine oil annually. Some whale species were driven to the brink of extinction. Each year, 15,000 right whales were slaughtered, and as these became scarce, the world's whaling fleet turned its attention to other species. Whales survive today only because the demand for whale oil plummeted.

In 1851, Abraham Gesner found that oil for illumination could be distilled from coal. He named it kerosene, from the Greek *keros* for "wax" and *elaion* for "oil" (*elaion* was altered to *ene* to sound like *camphene,* another lamp fuel). The new lamp illuminant was first produced in quantity in 1856. Just one year later, Michael Dietz, a

Brooklyn lamp merchant, introduced his clean-burning kerosene lamp to the public.

Kerosene burned cleanly, had no objectionable odor, and was much cheaper than whale oil. In addition, it could be stored indefinitely, unlike whale oil, which would eventually turn rancid. The obvious advantages overwhelming, the masses quickly turned to coal oil. Kerosene plants sprung into existence, with thirty operating in the United States by 1860, as whale oil was ultimately driven from the market.

While Dietz was busy selling lamps from his Brooklyn store and whale oil was offloaded from New England's whalers, Colonel Edwin L. Drake was drilling a hole in the ground near the small town of Titusville, Pennsylvania. Actually, Edwin Drake wasn't a colonel at all. He was a sometime railroad conductor who'd been laid off due to poor health and, until taking the drilling post, had been living with his daughter in New Haven, Connecticut. His current employers bestowed upon him his title, feeling Titusville's citizens would afford a colonel respect.

Titusville (pop. 125) was a logging town tucked into the mountains of northwestern Pennsylvania. But neither logs nor lumber brought Drake to this remote outpost—oil did. For centuries, the Seneca Indians had used the oil skimmed from the surface of creeks and springs near Titusville to make medicines. European settlers continued the Native American practice and dubbed the stuff "rock oil." They renamed a local stream Oil Creek and began pedaling its "magic liquid" as Seneca Oil. One vendor advertised its miraculous properties in a poem:

> *The Healthful balm, from Nature's secret spring,*
> *The bloom of health, and life, to man will bring;*
> *As from her depths the magic liquid flows,*
> *To calm our suffering, and assuage our woes.*

However, one person couldn't have cared less about Seneca Oil's curative properties: George Bissell, a New York lawyer who knew that rock oil was flammable. Bissell had a hunch that it might be possible to turn the stuff into an inexpensive liquid illuminant, and he soon convinced others to invest in this speculation. Among the investors was James Townsend, president of a New Haven bank.

For starters, Bissell engaged Professor Benjamin Silliman Jr., a Yale chemist, to evaluate rock oil as a precursor for both an illuminant and a lubricant. Silliman executed the necessary studies but withheld his final report until his hefty fee of $526.08 (about $13,000 today) was paid. After some contentious wrangling, Bissell raised the funds to pay Silliman and finally got his hands on the report.

It was everything that Bissell and his investors could have hoped for and represented the beginning of the modern petroleum business. Silliman found that the rock oil could be separated into several fractions via distillation. One of these fractions, he noted, was a high-quality illuminant. At the conclusion of his report, Silliman wrote:

> Gentlemen, it appears to me that there is much ground for encouragement in the belief that your Company have in their possession a raw material from which, by simple and not expensive processes, they may manufacture very valuable products.

Evaluation in hand, the initial investment group had no trouble rounding up other speculators, and stock was rapidly issued in the Pennsylvania Rock Oil Company. (Professor Silliman purchased two hundred shares.) The first challenge confronting the expanded investment group was proving that there was a sufficient supply of rock oil to support their planned industry (which many doubted). The second challenge was proving they could get at it.

There was good reason to believe that oil was a comparatively abundant resource. In biblical times, it was called bitumen, which oozes from the earth in various parts of the Middle East. Five thousand years ago, a bitumen industry was situated not far from modern-day Baghdad, providing great quantities of asphalt, which was traded throughout Mesopotamia as a mortar, waterproofing agent, and material for road construction. The walls of Jericho and Babylon were probably bound together with bitumen. It was likely the caulk used by Noah to seal the ark.

Oil was used as a weapon, too. In the *Iliad*, Homer wrote of a "flame that could not be quenched." Persia's King Cyrus was supplied with "pitch and tow" as he readied his attack on Babylon; it was to be used to set fires and force combatants from rooftops. Seventh-century Byzantines employed a mixture of petroleum and quicklime to make *oleum incendiarum*, or Greek fire; when exposed to moisture, it would ignite. They reduced attacking ships to ashes by hurling it onto the decks. In the mid-1800s, peasants in Galicia (parts of modern-day Poland, Austria, and Russia) "mined" oil through hand-dug shafts. Refined into kerosene, it became a staple of Viennese commerce. At the time Bissell was considering his next step, some estimates placed Galician crude production at 36,000 barrels annually.

Digging for black gold would impose the same production costs on petroleum-derived kerosene as that from coal. So, in 1856, Bissell was looking for an alternative to mining when he caught site of an advertisement promoting rock oil medications posted in a druggist's window. Next to the testimonials of the virtues of this medicine was a picture of several drilling rigs, like those used to drill for salt. Apparently, the rock oil for this medication was a by-product of salt production. The image triggered a eureka moment. What if Bissell *drilled* for the oil?

Within a month, jovial and friendly Edwin Drake, who just happened to reside, along with his daughter, at the same hotel as Bissell's partner James Townsend, was dispatched as "Colonel" Drake to Titusville. There he was to secure title to a prospective site where, the following spring, he would drill for oil.

Drake set up shop on the banks of Oil Creek, two miles downstream from Titusville. There he would spend the remainder of the year in a fruitless effort to hire salt borers. Renowned for their love of whiskey and drunkenness, Drake was cautious as to whom he hired. He sought to pay drillers in accordance with their production—one dollar per foot—but he found no one reliable to do the job. (The fact that many people in and around Titusville thought Drake insane probably didn't help.) As winter drew near, the colonel had nothing to show the suit-clad investors back in New Haven.

The following spring, fate smiled on Drake. He found a drilling crew, headed by "Uncle Billy" Smith, a blacksmith, and his two sons. The team built a derrick and started drilling. But progress was slow, and one by one Bissell's investors bailed. Finally, only Townsend remained, and when the last penny of venture capital had been expended, Bissell paid the bills out of his pocket. By the end of August 1859, even Townsend's enthusiasm and commitment had faded. Hopeless, he sent Drake a final payment and told him to settle his bills, close up shop, and return to New Haven. The letter arrived in Titusville on Monday, August 29, 1859.

Meanwhile, by Saturday, August 27, the crew had bored to 69 feet when the bit hit a crevice. Drilling was brought to a temporary halt. They decided to knock off for the remainder of the weekend and resume work on Monday. On Sunday, Uncle Billy visited the site to inspect the well and determine what had to be done to get the drill going again. When he looked into the drilling pipe, he saw a dark liquid.

The colonel arrived at the drill site on Monday to find Uncle Billy and his sons surrounded by an eclectic collection of containers, ranging from steel tubs and washbasins to whiskey barrels. Each one was filled with oil. With a simple hand pump, Drake spent the rest of the day extracting oil from the well. When he found Townsend's letter later, he ignored it.

Drake's achievement sparked a wild rush to Oil Creek. Worthless land was now priceless. Fortunes were made and lost. All of Titusville was awash with oil as the little logging town was transformed into the center of a new industry. Supplies soon outstripped demand, though, and prices plummeted, leaving whiskey barrels scarce and worth more than the oil they held. But the fortune made by Bissell, Townsend, and others motivated some to keep looking for oil.

The decade of the 1860s was marked by spectacular growth in all aspects of the new industry: exploration, refining, and distribution. Oil was discovered across the entire northwestern corner of Pennsylvania, earning it the nickname "Oil Region." Production soared from 450,000 barrels in 1860 to over three million barrels just three years later. Fifteen local refineries turned the crude into kerosene before it was shipped. Of the oil derricks and refineries of western Pennsylvania, one coal-oil producer remarked, "If this business succeeds, mine is ruined." A prescient insight, for by the end of 1860 kerosene from coal was a thing of the past. With bountiful supplies and abundant refining capacity, the most expensive aspect of kerosene production was distribution. In an attempt to break the teamsters' stranglehold on the cost of crude oil transportation, wooden pipelines were constructed. These proved to be more efficient and cheaper than horse-drawn transport, and by 1866 the oil of western Pennsylvania flowed directly to distant railroads via a network of pipelines.

Of course, this growth was fed by an insatiable demand for light. As Daniel Yergen notes in his book *The Prize,* less than a year after Drake's discovery, an oil handbook referred to kerosene's illumination as "the light of the ages," saying:

> Those that have not seen it burn may rest assured its light is no moonshine; but something nearer the clear, strong, brilliant light of day, to which darkness is no party. . . . Rock oil emits a dainty light; the brightest and yet the cheapest in the world; a

light fit for Kings and Royalists and not unsuitable for Republicans and Democrats.

It is hard to overstate the impact of kerosene on daily life, particularly on that of the rural population. In 1864, a New York chemist described the impact this way:

> Kerosene has, in one sense, increased the length of life among the agricultural population. . . . Those who, on account of dearness or inefficiency of whale oil were accustomed to go to bed soon after the sunset and spend almost half their time in sleep, now occupy a portion of the night in reading and other amusements; and this is more particularly true in the winter seasons.

To the business-minded, it seemed that oil was the foundation of a technological revolution, and at least one refiner believed that the route to unprecedented wealth and power lay in the new industry. The next chapter in oil's story picks up on a February day in 1865 when partners in one of Cleveland's most successful refineries were about to part company. The two had never agreed on how best to expand the business and had fallen into an argument, yet again, over the issue. Exasperated, Maurice Clark threatened to dissolve the business—a card he had played before. But this time the twenty-four-year-old John D. Rockefeller called his bluff. The two agreed to an immediate auction of the business, with Clark and Rockefeller the only bidders.

Clark opened at $500. Rockefeller countered. The bidding bounced back and forth this way until Clark reached $72,000. Rockefeller countered with a bid of $72,500, and Clark threw up his hands, saying, "The business is yours, John." They sealed the deal with a handshake.

Rockefeller knew that in the highly competitive kerosene market only the most efficient operations could weather the inevitable price fluctuations. Consequently, he set about making his business as efficient as possible. With borrowed money, he built the Standard Works, another Cleveland refinery. An export office was opened in New York, with his younger brother and partner, William Rockefeller, in charge. Eventually, exports would surpass domestic sales. Rockefeller took on another partner in 1867, a man by the name of Henry Flagler.

Rockefeller and Flagler became fast friends. They were completely absorbed by their business and continually planned its growth, their desks situated back to back. They designed and built refineries that were bigger and better than their competitors' and gained control of all the facilities required for kerosene production. They owned and operated a cooperage (barrel-making) plant, which was run with the same attention to efficiency as were their refineries, driving their cost for a barrel from three dollars to less than a dollar and a half. The partners manufactured the sulfuric acid used in refining and developed methods for its recovery and reuse. Their refineries were equipped with huge storage tanks for crude and refined oil. They also owned a drayage service, a New York warehouse, and boats to transport oil on the East and Hudson Rivers. They were the first to ship oil via tank cars, which—of course—they owned.

Their greatest contribution to the oil industry was transforming it into a petroleum industry. They did this by finding a use for *every* product distilled from crude oil: In addition to kerosene, Standard Works manufactured lubricants that were used in place of lard. Benzene was employed as a solvent and cleaning agent. Paraffin was used to make candles; naphtha, a dry-cleaning fluid; petrolatum, ointments; and white petrolatum became Vaseline. While other refiners thought of gasoline as a disposable by-product (often dumping it into nearby rivers where it occasionally caught fire), Rockefeller and Flagler marketed gasoline as a fuel.

Five years after buying out his first partner, Maurice Clark, John D.

Rockefeller re-formed his company as Standard Oil of Ohio, controlling 10 percent of the domestic oil business. Over the next forty years, the company grew into a behemoth, and in 1911, relying for the first time on the Sherman Antitrust Act, the U.S. Supreme Court ordered its dissolution. The Court found that the mega-company had acted to exclude others from the oil business. This was hardly surprising, as Standard transported more than 80 percent of all oil produced in Pennsylvania, Ohio, and Indiana. Seventy-five percent of U.S. crude was refined in Standard's refineries, and the company controlled more than 80 percent of the domestic and export kerosene market. When oil was transported over water, it was probably on one of Standard's ninety-seven ships. Railroads purchased more than 90 percent of their lubricants from Standard, and as for the railway tank cars, well, half of those were owned by Standard, too. But it wasn't just oil. Standard also produced a vast array of by-products, from millions of candles to tons of Vaseline.

When the dissolution of the company was over, Standard Oil became Standard Oil of New Jersey (eventually becoming Exxon), Standard Oil of New York (Mobil), Standard Oil (Chevron), Standard Oil of Ohio (Sohio), Standard Oil of Indiana (Amoco), Continental Oil (Conoco), and Atlantic (Sun).

While Standard indeed dominated the U.S. oil industry, it was not without domestic competition. In fact, if not for the whimsies of technology, Standard may very well have gone the way of the coal-oil business. The threat came from Thomas Alva Edison, prolific inventor of things electric and a marketing genius. In 1877, Edison turned his talents to the puzzle of electrical illumination, and two years later the incandescent lightbulb was ready for market—with Edison ready to found an industry to generate and distribute electricity. Conscious of his competition, Edison priced electricity competitively against town gas, the principal illuminant for city dwellers.

The new form of illumination had many benefits over gas and

kerosene. Its light was brilliant white. There was no odor and no flame to start fires. Where available, the public turned enthusiastically to electric light. Fewer than a quarter of a million bulbs glowed brightly in the homes and businesses of America by 1885, and by 1902, the number had grown nearly eighty-fold, to 18 million. Increasingly, kerosene was a "make-do" product restricted to rural areas where electricity was not yet available.

The oil industry's major market—illumination—was under attack, and there was little to be done to stave off the threat. Electric lighting was far and away a superior product. Still, there was reason for optimism: Entrepreneurs and inventors saw a future for the "horseless carriage." And when it came to powering these vehicles, oil had a distinct advantage over its competitors, steam and electricity. Internal combustion engines were more powerful than similarly sized steam engines, and they could run longer than electric motors powered by batteries.[5] The underappreciated component of oil that fueled the horseless carriage—gasoline—was now the hook upon which oil would hang its future.

One of the inventors and entrepreneurs who saw a future for internal combustion was the chief engineer at the Edison Illuminating Company in Detroit. Bitten by the horseless carriage bug, he quit his job and started designing and marketing gasoline-powered vehicles, which he would call Fords.

Henry Ford, and others like him, created a product that was so irresistible some might have even called it sexy. One observer fell just short of coining the early car a "chick magnet" when he wrote, "The man who owns a motorcar gets for himself, besides the joys of touring, the adulation of the walking crowd, and . . . is a god to the women." The growth of the automotive industry was as spectacular as that of the oil industry a generation earlier. Between 1900 and 1912, car registration increased more than a hundred times, from 8,000 to 902,000. Each of these cars consumed oil, and with more cars came a need for more gasoline. For the petroleum industry, what had once been a blessing was now a curse: Gasoline is but a minor constituent of crude oil.

Crude oil is a mixture of hydrocarbon molecules. You might think of these as strands of hairy pearls, where the pearls are carbon atoms and their hair is made of hydrogen atoms. Each pearl in the strand has two hairs, except the ones on the end, which have three. This molecular jewelry comes in different lengths, from a strand with one carbon pearl, which is methane, to strands with thirty or more carbon atoms. The number of carbon atoms controls the properties of the molecule. For example, if the strand has fewer than five carbon atoms, the molecule is a gas at room temperature; five to eighteen, a liquid; and nineteen or more, a solid. The most volatile components of crude are called naphthas, which are used as solvents and dry-cleaning fluids and are characterized by strands of five to seven carbon atoms. Gasoline is the component of crude with strings of 7 to 11 atoms in length. Kerosene is the portion containing strands of 12 to 15; diesel, 15 to 17; and lubricating oils are made of hydrocarbon chains of more than 17 carbon atoms. The 20-atom-plus range is where paraffin, tar, and asphalt are found.

Thus gasoline is one of the more volatile liquid components of oil, and it comes as no surprise that it represents but a small fraction of crude oil. Recall that short molecules will not form unless organic matter is buried deep in the oil window, and then, being more volatile, these will be the first lost to seepage or migration through the caprock. The problem facing the oil industry, and Standard in particular, was to find a way to wring more of their new moneymaker out of crude.

Before the 1911 breakup of Standard, William Merriam Burton, a Ph.D. chemist from Johns Hopkins and head of manufacturing at the Indiana refinery, had quietly directed his small research group to investigate ways to reap more gasoline from crude. Unbeknownst to everyone outside the Indiana plant, the research team decided to mimic nature's approach, cracking the crude into smaller chains by subjecting it to the great temperatures and pressures it would be exposed to when near the bottom of the oil window. This is fine when the pressure can be contained beneath several kilometers of rock, but in the lab this would be dangerous.

Refinery workers were understandably frightened and uncooperative, leaving Burton's scientists to run the hot, high-pressure stills in which nature was to be emulated. The experiment was successful: Gasoline production from a barrel of crude increased from natural levels of 15 to 18 percent up to 45 percent. But more important, Burton had transformed the petroleum industry into the petrochemical industry. It was now possible to engineer a process to provide from oil the molecules that one wanted.

There were also economic rewards for Standard (at least the part that would become Standard of Indiana). Prior to the court-ordered breakup, Burton had applied to headquarters for permission to build one hundred stills for gasoline production. He was refused, apparently because the process was potentially too dangerous. But after the dissolution, Standard Oil of Indiana approved the construction, and in early 1913 a synthetic gasoline known as "motor spirits" was sold in Indiana. The other offspring of Standard Oil were forced to license the lucrative Burton process from their sibling.

In the years following Colonel Drake's success, the lure of fast riches seeded oil exploration and kerosene production across great areas of the planet. In 1873, the land around Baku, on the western coast of the Caspian Sea, became the center of an Asian oil industry run by Robert and Ludvig Nobel, brothers of Alfred Nobel, the inventor of dynamite. Since the time of Marco Polo, the oil seeping to the surface here was renowned as good to burn and useful for cleaning mange from camels. By 1884, two hundred refineries in and around Baku were turning 10.8 million barrels of crude per year into the kerosene illuminating St. Petersburg's long winter nights.

At the turn of the nineteenth century, Sumatran oil enriched the coffers of the Royal Dutch Oil Company. Its genesis can be traced to a day in 1880 when a rainstorm forced Aeilko Jans Zijlker to take refuge for the night in a dark vacant tobacco shed. His guide lit a torch that burned with unusual brightness. Struck by the brilliant illumination, Zijlker questioned his guide and learned that the torch had been coated with wax skimmed from local ponds. Upon further investigation, he found that the ponds sat atop oil seeps that were almost 50 percent kerosene. Drilling rigs, pipelines, and refineries followed, and by 1897 the Dutch played a major role in global oil production.

In 1901, oil hunters drilled into Spindletop Hill in southeastern Texas. On January 10, their rig was blown to bits, as the pressure in the reservoir forced oil up the borehole at a rate of 75,000 barrels a day. This unprecedented phenomenon was christened a "gusher" and ignited the Texas oil boom. In 1903, 23 million barrels of crude flowed from California's oil fields; two years later, oil was discovered at Glenn Pool in Oklahoma. The oil world then shifted its attention to the Middle East in 1908 with the discovery of oil beneath Persian sands (in modern-day Iran).

Even in the face of the worldwide petroleum frenzy, oil was an economic rather than a strategic commodity. This, too, was about to change. As with the conversion from charcoal to coal, the British navy would play a leading role in oil's impact on global politics for the next hundred—or more—years.

The same year the Supreme Court ordered the dissolution of Standard Oil and drive-in service stations were popping up like buttercups along urban byways, Winston Churchill was appointed First Lord of the Admiralty. Germany's growing militarism had led him to conclude a European conflict was inevitable. Churchill's responsibility was to make the Royal Navy ready for the looming war. The future of the British Empire depended on his decisions, and primary among these was the question of coal or oil. Would Britain be better served by converting its navy to oil and internal combustion engines, or should they continue using coal and steam? There were powerful and persuasive arguments on both sides of this coin.

Internal combustion engines were more powerful, thereby making ships faster, and naval warfare was all about speed. The faster navy would be free to engage its enemy when and where it pleased. Churchill commissioned the Royal War College to determine how faster battleships might affect battle tactics. They found that a "Fast Division," with a top speed of 25 knots, compared with their then-current maximum of 21 knots, could outmaneuver and ultimately defeat the emerging German fleet. In addition, the conversion to oil would relieve manpower by eliminating the need for stokers.

The other issue was that of supply. Coal was plentiful in England. Welsh coal mines provided the necessary resources to keep the Royal Navy plying the waters of its global empire. But oil? None. Other countries that were converting their navies to oil (like the United States) would not be isolated from their required energy resources because they possessed large domestic supplies of oil. England's conversion to oil, however, would make its navy dependent on a continuing flow of foreign resources. Churchill decided that the stakes were too high not to convert. In April of 1912, he committed the Royal Navy to oil and to building the Fast Division, which would include five oil-powered Elizabeth-class battleships.

Solving the supply problem was now paramount. Churchill believed that the Admiralty should act to prevent the formation of a single universal oil monopoly, lest England fall dependent upon it. Early in 1914 prime minister Asquith agreed, and in a letter to King George V, he wrote that the government should "acquire a controlling interest in trustworthy sources of supply."

For the cash-strapped Anglo-Persian Oil Company, which was holding off takeover by Royal Dutch/Shell, the British government was a white knight. On June 17, 1914, Churchill spoke on the floor of the House of Commons. He proposed that the British government purchase a 51 percent share in Anglo-Persia (which in the years to come would morph into British Petroleum, BP) with a £2.2 million investment and place two representatives on the company's board of directors. In return, and among other provisions, the Admiralty was guaranteed an attractive fixed fuel price for two decades. Debate was intense, but the bill was finally approved. Europe was at war within months, and England's security was now intimately intertwined with the politics and economics of Middle Eastern oil.

The First World War, with the emergence of airplanes and mechanized infantry, served to drive home the point that combat had changed. Battles and wars now turned on oil's availability, transport, and distribution. Oil's new status as a strategic natural resource was all too obvious in the run-up to the Second World War. New resources discovered in Venezuela, Texas, and the Middle Eastern countries of Bahrain, Kuwait, and Saudi Arabia would figure large in the final disposition of this war. What ultimately drew the United States into World War II— the bombing of Pearl Harbor—had more to do with the oil fields of the East Indies.

Japan's imperial ambitions had been fanned by the hardships imposed on it by the Great Depression. The collapse of world trade had hit Japan particularly hard, and the resource-poor country was compelled to import a wide range of raw materials, including more than nine-tenths of its oil (80 percent of which was supplied by the United States). In the aftermath of the Depression, Tokyo began to articulate a policy of entitlement to the resources of Southeast Asia. At the same time, some elements of the Japanese military were planning for a modern war of conquest and expansion. Yet with the United States likely to oppose such an adventure, the Japanese had to find alternative oil supplies. Given the island nation's location, there was only one option: Sumatra.

Japan's aggression in China had begun with the disappearance of a Japanese soldier, who was "believed" to be in Chinese territory. Within three weeks of this incident Japanese forces had taken Beijing, and China and Japan were at war. The ongoing conflict had embroiled Tokyo and Washington in a game of chicken. As the two countries roared at each other, each pursued policies they hoped would make the other swerve. To conserve its resources, Japan initiated programs of rationing and synthetic oil production, and Tokyo politicians, wary of the American response, resisted Adolf Hitler's requests for more military and political cooperation. On the other side of the Pacific, export restrictions were placed on war goods, including high-octane aviation fuel. There were partial embargoes and hints, then threats, of total oil embargoes. Still, Washington was reluctant to take this step, recognizing that an embargo would likely precipitate Japan's move on the Dutch East Indies. The U.S. Pacific Fleet was deployed to Pearl Harbor to serve notice that the U.S. Navy would oppose any such action.

When, in the summer of 1941, the Japanese undertook an armed occupation of southern Indochina, President Franklin Delano Roosevelt was left with few options. He attempted to tighten oil exports without actually imposing an embargo; he was readying for war in Europe and did not want to initiate a second armed conflict with Japan. Roosevelt decided to roll back oil exports to 1935–36 levels and halt the export of products that could be used to manufacture aviation fuel. At the same time, he froze all Japanese financial assets in the United States.

Though the president had not wanted a total embargo, his assistant secretary of state for international affairs, Dean Acheson, did. And Acheson managed to get his way. In cooperation with the U.S. Treasury department, the frozen Japanese assets could not be used to buy

U.S. oil. From that point on (August 1941), U.S. oil was off-limits to the Japanese. Quickly, the British and Dutch followed suit, embargoing oil shipments from the East Indies. Japan was now isolated from the oil it needed to wage war.

In the year before the embargo, Japanese admiral Isoroku Yamamoto had proposed Operation Hawaii, a plan to attack the Americans at Pearl Harbor, incapacitating the Pacific Fleet long enough to secure the oil fields of Sumatra and Borneo. Given the oil embargo, the Japanese were out of options. Preparations for Operation Hawaii were accelerated.

On December 7, 1941, the Japanese began an all-out offensive. At the same time the USS *Arizona* was on its way to the bottom of Pearl Harbor carrying a human cargo of 1,177 sailors, Japanese bombs were falling on Hong Kong, Singapore, and the Philippines. This massive assault was not intended primarily as an attack on the United States, but as part of a coordinated campaign to capture the oil fields of the East Indies.

In the years following World War II, oil production in the Middle East soared. Israel declared its independence; construction of the U.S. interstate highway system began; and the first Holiday Inn opened (in Memphis, Tennessee). The Cold War heated up with the building of the Berlin Wall in 1961. Oil was discovered in the African countries of Algeria, Nigeria, and Libya. Iran, Iraq, Kuwait, Saudi Arabia, and Venezuela formed the Organization of Petroleum Exporting Countries (OPEC) to unify and coordinate members' petroleum policies. Oil was discovered on Alaska's North Slope and in the North Sea. And the United States went from being a net oil producer to a net oil consumer.

Then, in the early afternoon of October 6, 1973—Yom Kippur on the Jewish calendar—a combined assault from Egyptian and Syrian armed forces fell upon Israel. The Syrians attacked from the north with Russian-made jet fighters and seven hundred pieces of artillery.

More than two hundred Egyptian MiG aircraft screamed into the Sinai, bombing Israeli command positions while artillery pounded the east bank of the Suez Canal. The most potent weapon, though, had yet to be deployed. That would come in the weeks to follow.

Churchill had recognized that oil might be used to effect political ends: The United States had attempted to influence Japanese war planning with the threat of and later institution of an oil embargo. As ever more oil flowed from the Middle East to feed the world's energy hunger, the Arab nations considered using oil as a weapon to influence U.S. policy vis-à-vis Israel. With this intent, the Saudis cut exports during the 1967 Arab-Israeli conflict. However, the effects went unfelt as the United States upped its production to offset losses from the Middle East. In the end, the Saudis simply cost themselves oil revenues.

The situation had changed by 1973; U.S. oil fields were working at full capacity, and the escalating Arab demands for Washington to "soften" its support of Israel were taken seriously. The Nixon administration did what it could through that summer: engaged in a carefully choreographed dance designed to appease the Saudis while not abandoning the U.S. commitment to Israel. But the Yom Kippur invasion was not on the dance card.

The intensity of the conflict took the Israelis by complete surprise. Their two-week reserve of ordnance began to run low in just a few days; with ongoing resupply by the Soviet Union, the Arab combatants were unconstrained. Desperate, Israeli prime minister Golda Meir begged Washington for military equipment. The United States acquiesced and arranged for a C-5A transport to deliver munitions and other war matériel. The plan was to keep the resupply effort as low profile as possible, landing the aircraft under cover of darkness on Saturday, October 13, and departing as soon as it was unloaded. Unfortunately, weather delays brought the world's largest aircraft over Israel in Sunday's full daylight, its U.S. insignia shining bright.

Saudi Arabia took the public show of support for Israel as a slap in the face, and the situation got worse. On October 19, Nixon proposed a $2.2 billion aid package for Israel. Libya immediately cut all oil exports to the United States, and it took the Saudis only one day to make the same decision. The other Arab states followed in lockstep.

The consequences of the oil embargo were instantaneous. A barrel of oil, which months earlier had sold for a few dollars, jumped as high as $22.60. In the United States, where cheap gas seemed a birthright, motorists waited in line for hours to top off. Service stations often limited purchases to five or fewer gallons, causing drivers to go from station to station in an effort to fill their tanks. "No Gas" signs grew at filling stations, reminding Americans of our vulnerability and place in the new world order.

The Arab oil embargo produced shock waves that radiated through time. From 1973 onward, the world reacted rather than acted. In 1978, oil exports from Iran were halted as its shah was overthrown. Angry Iranian students took U.S. embassy personnel hostage. The price for a barrel of crude surpassed $30. Gas prices in the United States hit historic highs (in both real and nominal terms). Heating oil skyrocketed. Jimmy Carter donned a sweater and lowered the White House thermostat; Americans insulated their homes and drove smaller, "fuel-efficient" cars. Ronald Reagan was elected president in 1981 and promptly reset the White House thermostat back to a comfortable 76°F. Iraq declared war on Iran, and both sides pumped excess oil to fund their conflict. Oil prices collapsed in 1986. Two years later, Iran and Iraq agreed to a cease-fire. The Berlin Wall became meaningless on October 9, 1989, initiating the collapse of the Soviet Union.

With the end of the Cold War, new optimism arose over the prospects of a more peaceful world and predictable future. Yet within OPEC tempers ran high. Saddam Hussein accused Kuwait of driving oil prices down by violating its production quotas. The lower prices,

Hussein asserted, were costing Iraq billions of dollars in oil revenues. In reality, Hussein's rants were nothing more than a pretext for his ambitious plan to seize Kuwait. In July 1990, Hussein positioned 100,000 troops on the Kuwaiti border. Most observers saw this as an Iraqi show of force intended to prompt Kuwait to conform to its production quotas. The charade was exposed on August 2, when Hussein's forces invaded Kuwait.

The United Nations responded promptly to this naked aggression and as a way of frustrating Hussein's war plans imposed an embargo on Iraqi oil. Four million barrels per day were suddenly removed from the world oil supply. Once again, oil prices went through the roof and financial markets reacted with panic. For oil-consuming nations, the stakes were monumental. If Iraq was going to control Kuwait's oil, Hussein would control 20 percent of OPEC production and 25 percent of the world's oil reserves. President George H. W. Bush and the leaders from a coalition of nations reacted accordingly, driving Hussein's forces from Kuwait. Iraq was hobbled. The coalition troops enforced no-fly zones in both the north and south of Iraq, and the UN imposed restrictions on the country's oil exports and forced it to submit to arms inspections.

Oil prices fell steadily through the 1990s, reaching historic lows in real terms. Economic expansion and prosperity swept the United States: Oil's availability no longer seemed to concern Americans as the SUV became the most popular vehicle to crawl through rush-hour traffic. The demand for oil in developing countries, particularly India and China, was on a steady climb. The world demand was rapidly approaching oil's supply capacity, but it went virtually unnoticed, at least by the public.

In March 2003, President George W. Bush ordered U.S. troops to invade Iraq. The Bush administration asserted that the invasion was necessary to prevent Saddam Hussein from continuing to accumulate, and possibly use, weapons of mass destruction. The ground offensive was quick and decisive. Coalition troops occupied Baghdad on April 9, 2003, and in the days following the fall of Baghdad, looting was widespread, including the National Museum of Iraq. But Iraq's oil fields and its oil ministry were safe. They were secured by U.S. troops.

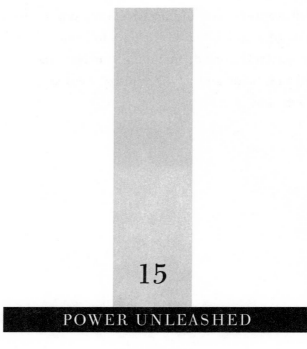

15

POWER UNLEASHED

Electricity

C oal fueling steam engines and oil fueling internal combustion engines brought the twentieth century into rough relief. Electricity
is the instrument used for finishing and polishing. Electricity liberates
energy from its source. No longer did each human brain need its own
steam engine or water wheel to fuel its imagination, just a wire carrying
electric current. Suddenly everyone had access to the energy his or her
mind craved. And what was more important, from the perspective of
the energy diet, it didn't matter how the electricity was generated.
Whether it was produced by burning coal, natural gas, or in a hydroelectric dam, electricity is electricity. It is the most versatile form of
energy.

For years, Michael Faraday searched for the invisible forces responsible for electricity and magnetism. And for years many of his peers considered his ideas absurd. How could there be anything in empty space? Wasn't empty space just that, empty? But Faraday had come up with a very different picture to explain why magnets attracted pieces of iron. To him, the space around a magnet was filled with lines of force that pushed other stuff around (like iron). As the English chemist and physicist approached forty, he was on the heels of a discovery that would forever change the way we use energy.

Faraday knew that when electricity—electrical current—moved through a coil of wire wrapped about a piece of iron that the metal became a magnet. But in 1831 Faraday was pondering the opposite question: What if a magnet were to move *inside* a coil of wire? He hooked his wire to a meter used to detect current, then moved a bar magnet back and forth within the wire's coils. The needle on the meter moved. In one step, Faraday had invented the electric motor and the electric generator.

From a modern perspective, the principles behind electricity generation are not particularly hard to understand. We know that an electric current is the result of electrons moving down a wire, more or less together. The best way I know to visualize this motion is through a thought experiment. Imagine a wasp's nest, one of those shaped like an upside-down pineapple. Wasps are flying all around the nest. Some are walking about on its surface, and still others are inside, doing whatever wasps do. Now gingerly take the nest in your hands, and so as not to disturb its residents walk slowly along the street. If you are careful—in a thought experiment you can be as careful as you wish—the wasps will move with you. They will still circle their nest and walk about on its gray papery surface, but overall they will move as you do. The wasps will "drift" along the street.

The motion of electrons in a wire is analogous to the motion of our imaginary wasps. When there is no current (the stationary nest), the electrons zip about in their orbitals but don't move along the wire. When the current is turned on (the nest starts to move), the electrons drift down the wire. This "electron drift" is electric current.

In addition to having an electric charge, electrons are also little magnets, and moving magnets exert a force on these electrons, causing them to drift. That is, a magnet moving relative to a wire will induce an electric current. The phenomenon is called *magnetic induction.* But remember that electrons are magnets, too, so an electric current also exerts a force on a magnet and causes it to move. This would prove invaluable, at least when it came to using energy.

The problem with steam and internal combustion engines is that they are big, noisy, heavy, dangerous, smelly, and require constant attention. They are not particularly practical when it comes to doing odd jobs, like running a fan or powering a toothbrush. Think about it: You get up in the morning and have to stoke your toothbrush to build up a head of steam before you can even start to clean your teeth. Then there is the problem of what to do with the waste steam and coal exhaust. And where will the coal be stored, in a tube next to the toothpaste? But an electric toothbrush, that's another matter. At one end is a generator— a magnet that rotates within a coil of wire; at the other end is an electric motor—a magnet that can rotate in response to the wire's current. Connect the generator end to an energy supply, say, a waterwheel or steam engine that turns the magnet, and a current starts to flow in the wire. Almost instantly, the magnet at the other end begins to spin. The magnet is attached to the head of our toothbrush and, voilà, plaque is a thing of the past! In effect, the power of the waterwheel or steam engine has been transmitted down a conduit to where we can use it, and all the while we remain oblivious to the noise, smell, and bother going on at the other end.

The first revolutionary change to follow from Faraday's insights was the telegraph. The idea was so obvious that inventors in America and England seized upon it almost simultaneously. In America, credit goes to Samuel Morse and Alfred Vail; in England, Charles Wheatstone and William Cook made the discovery.

The pairs of scientists used a battery to power an electric circuit. In its simplest form, at both the sending and receiving ends, the circuit ended in a coil of wire wrapped around a small cylinder of iron. When the transmitting lever was depressed at the sending end, the circuit was completed and a current appeared in the coil at the receiving end. The iron cylinder became a magnet and attracted a lever. The exact sequence tapped out at the transmitting lever was reproduced at the receiver. Using Morse code, messages could be transmitted over long distances almost instantaneously.

Telegraphy alone could have guaranteed electricity a prominent spot in the energy hall of fame, but electricity is versatile. Among other things, it's what powered Edison's incandescent lightbulb. To supply energy to his invention, Edison built his first commercial power station on Pearl Street in Lower Manhattan in 1882. A coal-burning steam engine ran a generator that distributed electricity to fifty-nine customers over one square mile. Initially, the electricity from this station was only used to power the "Edison Electric Light Bulb." As such, the power station operated at full capacity during the time lightbulbs were on—after dark.

Electric motors delivered the power of steam or internal combustion to the masses with none of the disadvantages (if we don't look beyond the walls of our house or factory, at least). Thus, within a few years of the Pearl Street plant's opening, demand for electricity had increased so dramatically that it was forced to operate twenty-four hours a day. Because the electricity generated at Pearl Street was direct current (DC), it could be transmitted only a few blocks. By the end of the 1880s, the demand for electric power had created a proliferation of local generating stations, which dotted many U.S. cities.

The technology to transmit electrical current over long distances came in 1887 when Nikola Tesla introduced generators, motors, transformers, and lights that used alternating current, or AC. The advantage lies in the ease with which AC can be transformed to higher and lower voltages.

The voltage, current, and resistance of an electric circuit control its power. To get a better feeling for these quantities, we can draw an analogy between electrons drifting down a wire and water flowing in a river. The river's power can be quantified if we know how much water the river holds and how fast it is moving. For example, the water in the mighty Mississippi moves slowly, but there is a lot of it, so it is a powerful river. On the other hand, a stream plummeting down a steep mountain slope is also powerful, for although there is not much water, it is moving quite fast. The amount of flowing water is analogous to a circuit's current, its velocity to the circuit's voltage. The energy carried by the circuit, like a river's power, is determined by a combination of the two. To restate the wasp's nest analogy, a circuit's voltage is given by the electron drift velocity (how fast you are moving with the wasp's nest) and its current by the number of electrons that drift past a point in a given time (the number of wasps in the nest). Finally, anything that impedes water flow contributes to resistance. Log jams and whitewater-producing boulders reduce both a river's flow and velocity, increasing its resistance. Resistance to electrical current is controlled by a wire's diameter and by the kind and arrangement of atoms from which it is made. For example, copper wires have a lower resistance to electrical flow than do similarly sized ones made of aluminum.

The Second Law of Thermodynamics mandates energy losses for any normal circuit. However, increasing voltage and decreasing current minimize these losses. So if you want to transmit energy over great distances with minimal loss, it is best to use high voltage. But now we have another problem: High-voltage electrons can jump across large gaps (this is why Taser stun guns are high-voltage instruments). High-voltage current is not something you want in your home.

The solution is an obvious one: Transmit electrical energy using high voltages and low current. When it gets to where it is going, switch to a low voltage and high current. In terms of our water parallel, this would be like moving water from its source to our homes using high-pressure hoses, which would fill a large cistern on the roof. The fast-moving water from the hose is equivalent to high-voltage electricity. The water flowing into the house from the cistern won't move very fast, but with a large enough drain pipe, a lot of water can be made to flow (low voltage, high current).

The key here is that efficient and safe transmission of electricity requires a means to transform power between different voltages. For AC, this is a simple matter; for DC, it is not. Hence, alternating current became the standard. In the United States, AC up to 765,000 volts flows through transmission lines before it is transformed into the 110 volts used by blenders, toasters, lightbulbs, and fans.

Once current could be transmitted over distance, it was feasible to harness energy in one place and use it in another. And one place where there was an excess of energy waiting to be harnessed was Niagara Falls. I once had an occasion to spend the night in Niagara Falls when I was driving from Boston to Denver. Believe it or not, the shortest route between these two cities is to cross into Canada at Niagara Falls, New York, and then back into the States at Detroit. What I remember most about my evening was the roar.

The falls' tremendous growl comes from a voice box powered each second by 40,000 metric tons of water rushing over its edge to plummet roughly 50 meters and explode against the rocks and river waiting below. The total energy content of Niagara is astounding, generating on the order of 20 billion watts (20 gigawatts, denoted GWe).

The first known person to tap this energy was Daniel Joncairs. In 1759 he diverted water from the American side of the river through a hand-excavated ditch. The flow was sufficient to turn a waterwheel that then powered Joncairs's small sawmill. However, others saw in his ditch the potential for something a little bit bigger.

The plan was to construct a canal 70 feet wide, 10 feet deep, and 4,500 feet long that would draw water from the Niagara River a half-mile upstream from the falls. It would then channel the water to a point below the falls but 210 feet above the river. There it would be discharged through water turbines. The resulting mechanical power would then be transmitted via belts and drive shafts to the factories above. Construction began in 1853. It would take more than twenty-five years to complete, leaving several visionaries bankrupt. However, by 1881, then-owner Jacob Schoellkopf watched with satisfaction as his man-made waterfall provided hydropower for the Niagara Falls flour mills. In the years between the inception and completion of Niagara's hydraulic canal, Edison had invented incandescent light-bulbs. The demand for electricity was growing and Schoellkopf saw a future in providing for this demand.

It was about this time that Charles Brush appeared on the scene. Brush was an inventor from Ohio who wished to showcase electric carbon arc lamps by illuminating Niagara Falls at night. Schoellkopf volunteered his water turbines to power an electric generator, and in 1881 arc lamps bathed the falls in brilliant light. The following year, Schoellkopf teamed up with the Brush Electric Light Company to construct a generating station at the end of his canal. The electricity generated there supplied sixteen Niagara Falls streetlights.

Thomas Evershed, a New York State engineer, dreamed big. Power to run flour mills, light streets, and show off Niagara Falls hadn't made a dent in the tremendous energy available from the Niagara River. He envisioned a 200,000-horsepower generating station that would utilize the falls' full height and linked canals and discharge tunnels buried beneath the town of Niagara. Not a bad idea, if all the electricity could be used. But more power than the local mills could possibly use would result. Without an effective means to transmit the extra electricity, the canals, tunnels, and generating stations didn't make economic sense.

There was, of course, a solution to the electric distribution predicament in Nikola Tesla's alternating current. The American inventor, manufacturer, and thorn in Edison's side George Westinghouse knew all about Tesla's work and in 1883 designed an AC illumination system for Niagara Falls. With alternating current, Niagara's moneymaking potential was plain to see. That's what prompted J. P. Morgan, John Astor, and William Vanderbilt to take an interest in the waterfall.

With venture capital from some of the wealthiest men in America, The Niagara Falls Power Company built a tunnel 49 meters (160 ft) underground that was 2,042 m (6,700 ft) long, 6 m (21 ft) high, and 5 m (18 ft) wide. Upstream was Powerhouse #1, a 140 m (450 ft) building housing the generators that could churn out over 100,000 horsepower of AC.

A year after the completion of Powerhouse #1, Westinghouse Company worked with Niagara Falls Power Company to develop and build a long-distance electric transmission grid that would include 11,000-volt transmission lines. On November 15, 1896, electricity flowed to the city of Buffalo along these wires. Good use was made of this inexpensive electrical power, and Buffalo became the first American city to be lit by electricity, earning it the nickname "City of Lights." But the lights did more than brighten dark nights; they also served as welcoming beacons to the energy-intensive industries that flocked to the Buffalo-Niagara region.

First on the scene were the electrochemical industries. They came for the bountiful energy supply, but specifically for the electricity, which was used straight out of the wire to separate elemental metals from ores and compounds. The Pittsburgh Reduction Company—a producer of aluminum metal—opened its first Niagara plant in 1895 and a second a year later. The company reorganized in 1907 as the Aluminum Company of America (ALCOA); in 1914, ALCOA used more hydroelectric power than any other company in the world. At the turn of the century, most of the world's aluminum was produced in the United States, and most of that came from Niagara Falls. The Niagara Electro Chemical Company opened a plant on the banks of the Niagara River in 1896, where the inexpensive hydroelectric power reduced the costs to manufacture chlorine and metallic sodium (Du Pont eventually absorbed that company). The Titanium Alloy Manufacturing Company, which used electrolytic processes to produce titanium metals and alloys for steel making, also moved to Niagara in 1906.

Electrical power offered a clean energy source for baking. Henry D. Perky relocated his Natural Food Company from Worcester, Massachusetts, to its new home, an all-electric model factory in Niagara Falls in 1901. Two of Natural Food's best-known products were Shredded Wheat breakfast cereal and Triscuits—wheat wafers baked in electric ovens. Perky marketed his product as a "Wonder of the Age" and printed an image of the falls on each box of Shredded Wheat. In 1919, the National Biscuit Company (NABISCO) purchased Perky's company.

As Niagara and Buffalo grew, so did the electrical power industry: Powerhouse #2 followed Niagara's Powerhouse #1. On the other side of the river, generating stations were popping up to supply Canada with electricity, and transmission lines spread like spiderwebs across New York and Ontario. Sixty 1,000-volt transmission lines connected the city of Toronto to generators pulling power from the Niagara River. So much water was being diverted for power generation that, by 1909, the falls were in danger. The United States and Canada entered into a treaty (one that has been amended numerous times) to guarantee minimum water flows over the falls. At the same time, Niagara's maximum generating capacity—roughly 4.5 billion watts—was also pegged.

Even with so much power contained in Niagara Falls, it was far from enough to meet the electricity cravings of the United States in the early twentieth century. But there weren't many massive waterfalls to be tapped. The next best thing was a dam spanning a powerful river, so the Pacific Northwest soon became the most hydroelectrically developed region in the world. With eleven major dams on the Columbia River and hundreds on its tributaries, 21 billion watts of high-voltage electrical power flowed along hundreds of miles of transmission lines. The first major hydroelectric producer on the Columbia was the Rock Island Dam, which was completed in 1932; Colossal Bonneville and Grand Coulee were completed in 1938 and 1941, respectively. Dam building on the Columbia finally ended in the 1970s as the last bit of economically feasible power was squeezed from the river.

The American Southwest answered when opportunity knocked as well. The Hoover Dam, built during the Great Depression, was the largest of its time. It turned the seasonal flow of the Colorado River into a reliable power source. Supplied by the 247-square-mile Lake Mead, the generators deep within Hoover Dam churn out 2 billion watts (roughly equivalent to the power generated by three million horses). Arizona's Glen Canyon Dam, with a generating capacity of 1.3 billion watts, took seven years to build; construction was completed in 1963 but it took another fourteen years to fill Lake Powell.

By the middle of the twentieth century, electricity wasn't just an energy source, it was what made things work. Televisions, radios, coffeemakers, hair dryers, washer-dryers—all were useless without electricity. And hydroelectricity provided only a small fraction of the power needed to keep all these gadgets running. At that time, most of America's electricity was generated by burning coal to make steam that turned a steam turbine that rotated a magnet that made electric power. Coal, of course, wasn't free—as gravity was—so coal-produced electricity was more expensive than hydroelectric. However, another energy source could be used to make steam to drive those same generators. Many thought it would make electricity even more economical than hydropower.

———

R ecall our energy ride. The lowest energy point in Element Range was Iron Valley. This means that fusing nuclei to form atoms lighter than iron liberates energy, while fusion to make elements heavier than iron absorbs energy. In the reverse of fusion, heavy elements can liberate energy by splitting into two or more smaller atoms. We call this process *fission*. A massive nucleus breaks apart into pieces, which combined weigh less than when they started. But the lost mass isn't lost at all; it has been converted to energy—what we call nuclear energy.

Fission is a common process. It goes on all around us. It is through fission that radon gas may accumulate in the basements of our homes. As a kid, I remember looking across the audience in a movie theater, and from almost every wrist there emanated the unnatural green glow of radium fission. The hands and numbers of every wristwatch were painted with a compound containing the element radium. Up until the late 1960s this paint was also used to illuminate bedside clocks as well as many marine instruments. Radium spontaneously undergoes fission, releasing energy that interacts with other compounds in the paint to produce the strange green light. Fortunately, radiation exposure from all of these sources was not large enough to affect the greater population. However, for those workers who actually applied the paint it was another matter.

In Switzerland, the home of fine timepieces, it was common for the citizens of Zurich to say "there were so many radium painters . . . that it wasn't unusual to recognize them on the streets even on the darkest nights because of the glow. . . . Their hair sparkled almost like a halo." The most extreme consequences from radium exposure came not through skin contact but through ingestion! Today this practice would precipitate multiple OSHA violations, but for the dial painters of the 1920s it was routine. Robley Evans, an MIT professor and founder of the field of nuclear medicine, described the practice:

In painting the numerals on a fine watch, for example, an effort to duplicate the shaded script numeral of a professional penman was made. The 2, 3, 6, and 8 were hardest to make correctly, for the fine lines which contrast with the heavy strokes in these numerals were usually too broad, even with the use of the finest clipped brushes. To rectify these broad parts the brush was cleaned and then drawn along the line like an eraser to remove the excess paint. For wiping and tipping the brush the workers found that that either a cloth or their fingers were too harsh, but by wiping the brush clean between their lips the proper erasing point could be obtained. This led to the so-called practice of tipping or pointing the brush in the lips. In some plants the brush was also tipped before painting a numeral. The paint so wiped off the brush was swallowed.

By 1926, brush tipping was a lost art, spurred into extinction by the deaths (and lawsuits) of dial painters in the mid-1920s.

Although fission can be used to provide energy, it has one big drawback—it's uncontrollable. Let's say we have a kilogram or so of radium. At any one time, some fixed and very small fraction of the radium atoms are splitting apart and releasing energy. Hence the amount of energy generated by this one-kilogram mass is very nearly constant. It could be used to make steam, which in turn could power an electric generator. But what do we do on a hot day in Los Angeles when everyone turns on air conditioners? There is no way to demand more power from radium. Alternatively, what do we do on a beautiful, sunny spring afternoon? The temperature is perfect, no one is working or doing chores, and all the streetlights are off. This time, there is no way to throttle back the energy production. Electricity will be wasted—turned to heat and dissipated as entropy. If fission is to be an efficient source of electrical power, we need to vary its rate of energy production.

For that, a special fissionable substance is necessary. We need a fissile material.

It is a matter of geopolitical good fortune that fissile isotopes are scarce, for this is the stuff of which atomic bombs are made—for example, uranium-235 and plutonium-239. A fissile material is one that can be induced to fission. Just like radium, uranium-235 (92 protons and 143 neutrons) splits apart and releases energy at a constant rate, but it can be made to fall apart quickly by smacking it with a neutron. When this happens, not only does U-235 release energy, but it also kicks out a couple more neutrons. In turn, these smack into other U-235 nuclei, they split, release energy, and make more neutrons. It's not long before there are a lot of neutrons smacking into U-235 and making energy, which is called a *chain reaction.*

The bombs that were dropped on Hiroshima and Nagasaki were uncontrolled chain reactions. This is fine for a bomb, but it's not very practical when it comes to making electricity, where the chain reaction must be slowed down a bit. The trick to throttling back this reaction is to limit the number of neutrons zipping around, which is achieved with the aid of a neutron-absorbing "control rod" made of, for example, boron.

The principle behind the design of a nuclear reactor is elegantly simple. A core of fissile material is interlaced with control rods. To increase the fission rate, the rods are withdrawn from the nuclear core. More neutrons are then busy inducing fission. By advancing the rods into the core, the fission slows as neutrons are absorbed before smacking into a fissile nucleus. In summary, control rods *out* mean more neutrons, more fission, more heat; control rods *in* mean fewer neutrons, fission slows, less heat.

Turning the heat of fission into electricity is achieved with a coolant circulating about the core. Usually the coolant is water, but it can be a gas or even a molten metal that picks up heat from the core. Then, fission's energy is transferred to water in a heat exchanger. The coolant makes its way back to the core for another load of heat while the water—now steam—powers a turbine driving an electric generator.

The core of a nuclear reactor contains only a small amount of fissile material. U-235 constitutes about 0.71 percent of naturally occurring uranium. The remainder is mostly U-238, which is not fissile, though it does fission at a constant, though slow, rate. To make "reactor-grade" uranium, the U-235 must be enriched to about 3 percent of the core composition. "Weapons-grade" uranium or plutonium is almost pure U-235 or Pu-239.

Fortunately, isolating U-235 is an expensive, time-consuming, and technically demanding process. It is not the design or construction of an atom bomb that presents the obstacle; rather, it is obtaining the fissile material. There are two ways to do this: One involves processing great quantities of uranium ore to isolate the U-235, a relatively conspicuous activity; the other is to reprocess nuclear reactor fuel.

About 97 percent of a reactor's core is U-238. Though the nucleus of this atom will not spontaneously split apart when hit by a neutron, there is another effect. When U-238 absorbs a neutron, it becomes U-239, which is unstable and decays to Pu-239. This is a fissile material and is used in weapons. So while a reactor is running, fissile plutonium is being produced.

Countries with nuclear ambitions can maintain a somewhat lower profile by building a nuclear reactor to "meet their energy needs," so to speak, and every few months shut it down to extract plutonium from the reactor fuel rods. It is the job of the United Nations Atomic Energy Commission to inspect reactors to verify that they are operating as electric power plants and *not* as plutonium production facilities. The key indicator is the length of time between refueling. Short cycles mean someone's making plutonium.

Though a nuclear reactor cannot blow up like a bomb, it can melt down. The temperature of an uncooled core of U-235 and U-238 can rapidly exceed the metal's melting point, transforming it into a molten pool of radiation. Against this possibility, the core of most commercial reactors is situated in a structure of concrete and steel called the containment. The name obviously reflects its purpose to "contain" core meltdowns.

The first man-made nuclear reactors were built as plutonium production facilities during World War II.[6] The British built the first commercial nuclear power station at Calder Hall in the late 1950s. Generating 200 million watts, the plant operated for forty-seven years, closing on March 31, 2003. Through most of its service, it was used to produce plutonium, with electricity generation as a secondary purpose. The first full-scale U.S. power plant was the Shippingport reactor, located on the Ohio River about 25 miles from Pittsburgh. Originally designed to power a large aircraft carrier, the reactor was later adapted for commercial use, which included the production of plutonium. It operated for twenty-five years until October 1, 1982. Six years later, the 956-ton (870-T) reactor core assembly was lifted out of the containment building and shipped to a burial facility in Washington State.

The years between 1960 and 1990 marked rapid expansion of nuclear power facilities, with just under four hundred plants operating worldwide. The French had the most aggressive reactor program, generating almost 80 percent of their electrical power with nuclear energy. In comparison, nuclear reactors generated about 20 percent of the electrical power used in the United States. The rate of reactor construction gradually slowed and then ground to a halt in 1989 with the accident at Chernobyl. But even before that catastrophe, accidents had tarnished the bright future of nuclear energy.

In 1957, one of two plutonium-producing military reactors at Britain's Windscale nuclear site caught fire, releasing radioactive isotopes into the environment. Though no one was injured, the risk of cancer to the local population was deemed to have increased substantially. The accident at Idaho's SL-1 reactor was more spectacular and the consequences more evident. At 9:01 P.M. on January 3, 1961, during maintenance procedures following an eleven-day holiday shutdown, the SL-1 went "prompt critical"—a rapid increase in the number of fission events. In four milliseconds, the heat generated by the resulting power surge explosively vaporized the core's water coolant, propelling the reactor vessel upward. The operator, standing on its top, was killed instantly when he was crushed against the containment ceiling.

The worst pre-Chernobyl nuclear accident occurred at the Three Mile Island Generating Station on the Susquehanna River near Harrisburg, Pennsylvania. The incident at Three Mile Island's reactor number 2 began at 4:00 A.M. on the morning of March 28, 1979. For reasons still unknown, the plant's main pumps supplying water to the nonnuclear cooling system failed, causing the turbines and nuclear reactor to shut down automatically. This initiated a series of mechanical failures, miscalculations, and oversights that two hours and ten minutes later left the top half of the reactor core without coolant. The extreme heat drove a chemical reaction that produced hydrogen and other radioactive gases. At the same time, radioactive coolant began to leak into the containment building and the plant's primary coolant water. This went undetected until radiation alarms sounded two hours and forty-five minutes into the accident. By now, radiation levels of the coolant water were three hundred times higher than normal and the nuclear facility was seriously contaminated.

A general emergency was declared at 7:24 A.M. Harrisburg radio station WKBO announced a problem with the plant at 8:25 A.M. The Associated Press picked up the story at 9:00 A.M.

It would take almost sixteen hours to bring the reactor under control. In the interim, the hydrogen within the reactor ignited and burned, and a large part of the core melted but did not penetrate the reactor vessel—let alone the containment building. Over the next week, some direct venting of radioactive gas and steam to the atmosphere was authorized, though not without controversy. According to the American Nuclear Society, the net effect of the Three Mile Island radiation release was to expose the people living within ten miles of the plant to no more than 100 millirems of radiation. This is about one-third of the yearly background radiation absorbed by an average person.

An effective containment building mitigated the accident at Three Mile Island. In contrast, the Chernobyl nuclear reactor built without total containment was a disaster waiting to happen.

The world's worst reactor accident occurred in northern Ukraine on April 26, 1986, at the end of an experiment to test the reactor's safety features in the event of a power failure. During the test, operators switched off numerous safety features in direct contravention of the reactor's operational guidelines. Included in the list of violations was the withdrawal of at least 204 of the reactor's 211 control rods. Guidelines called for a minimum of 15 to remain in place during operation.

The control rods of the Chernobyl reactor were improperly designed and when first inserted caused an acceleration of the chain reaction. When all 204 rods were introduced into the core, the reactor's power output jumped to nearly 30 billion watts—ten times its normal output. The temperature of the core climbed far beyond safe limits and the fuel rods began to melt. Water turned explosively to steam, expanded vertically along the reactor's rod channels, and blew a hole in the roof.

In the absence of a containment building, the radioactive contaminants were released to the atmosphere. To make matters worse, high temperatures inside the reactor core sparked a fire that was fed by in-rushing air. The soot and smoke rising from this fire transported additional radioactive material high above Chernobyl. There it was caught by the wind and carried westward across Europe.

Among the residue of the nuclear meltdown and fire were particles of nuclear fuel and radioactive isotopes including cesium-137, iodine-135, and strontium-90, which were even more dangerous. The residents of the surrounding area observed that the cloud above Chernobyl glowed on the night of the explosion.

Thirty-six hours after the accident, Soviet authorities began evacuating the area around Chernobyl's reactor. Within a month, the 116,000 residents within 30 kilometers (18 miles) of the plant had been relocated. But radiation effects reached farther than that. The first the West knew of the accident was when Sweden's radiation detectors sounded their alarm. For months, satellites tracked Chernobyl's deadly cloud as it rained nuclear fallout about the globe. The disaster touched the entire Northern Hemisphere and awakened fears of the nuclear age.

Combined, Three Mile Island and Chernobyl cast a chill over the nuclear industry. Construction in the United States ceased, and some power plants were left half-completed. The belief in nuclear power as a provider of bountiful and inexpensive electricity vanished under the harsh light of experience.

Our Energy Present

16

FIRST, THE BAD NEWS

It is almost impossible to get through a day without being reminded of the dark clouds on the energy horizon. The dual thunderheads of failing oil supplies and global climate change cast a pall on the future. At times it appears the storm is already upon us. The surge in oil costs is blamed not on an embargo, but on global demand surpassing supply. From now on, the thinking goes, it is all downhill for production and uphill for prices. The consequences of ever-escalating oil prices are legion, ranging from predictions of world financial collapse to wide-scale civil unrest.

Scared yet? If not, then consider what some say about global climate change, or the more familiar "global warming." Indeed, some forecasts are dire: The ice caps over Greenland and much of the South Pole will melt in a matter of years and sea level will rise 6 to 15 meters (20 to 50 feet). Florida will disappear beneath the waves, as will many inhabited islands of the Pacific and Indian Oceans. Hundreds of millions, if not billions, of people will be displaced, their homes, villages, and cities flooded. The melting of Greenland's ice cap will reduce the salinity of the North Atlantic and Arctic Oceans, bringing the Global Ocean Conveyor (discussed in chapter 10) to a standstill. Europe will be plunged into an ice age. Mid-latitudes will experience more extreme weather, ranging from drought to deluges. Grain production will crash and famine will sweep the earth.

These predictions, whether accurate or not, are probably intended to scare us into action. Fear, particularly on this scale, immobilizes, or worse yet, prompts denial. Hence rather than look at the world as it may be, let's look at the world as it is. To most of us, this is less terrifying; the world of today seems more manageable—simply because we're managing.

In today's world, oil production stands near 80 million barrels per day (bpd). Of this, the people of the United States consume 25 percent. Each day we use 9.2 million barrels of gasoline and about 7.4 million barrels of other kinds of fuel oil (diesel, jet fuel, marine fuel). That means that around one-fifth of the world's oil production is used to move around 5 percent of its inhabitants and their belongings and trash.

So while we use more than 16 bpd for transport, we use another 5 bpd for electric generation, chemicals, etc. We consume more than 20 million barrels of oil a day, but we produce less than 9 million bpd. We import the rest. In May 2006, as the national trade deficit hit $63.8 billion for the month, we wrapped up $27.9 billion and sent it overseas to pay for 360 million barrels of black crude. In other words, oil imports will account for about 45 percent of our estimated record 2006 trade deficit ($763 billion).

Reflecting on the disparity between the amount of oil the people of the United States use compared with everyone else, a reporter asked Bush administration press secretary Ari Fleischer the following question at a May 7, 2001, Washington press briefing:

> Does the president believe that, given the amount of energy Americans consume per capita, how much it exceeds any other citizen in any other country in the world, does the president believe we need to correct our lifestyles to address the energy problem?

Fleischer's answer:

> That's a big no. The president believes that it's an American way of life, and that it should be the goal of policymakers to protect the American way of life. The American way of life is a blessed one. And we have a bounty of resources in this country. What we need to do is make certain that we're able to get those resources in an efficient way, in a way that also emphasizes protecting the environment and conservation, into the hands of consumers so they can make the choices that they want to make as they live their lives day to day.

Five countries provide about 70 percent of the imported oil necessary to provide America with its blessed way of life. Canada is the principal supplier, with Saudi Arabia, Mexico, Nigeria, and Venezuela bunched together in a virtual tie for second place. Ironically, some of the money we expend to maintain our "rights" to devour energy finds its way into the hands of those teaching radical anti-Americanism (specifically, the Wahhabi madrassas, or Muslim schools, sponsored by the Saudi government).

Saudi Arabia sits atop the world's largest proven oil reserve—

267 billion barrels—of which it pumps roughly 11 million bpd. In 2005, Saudi oil revenues topped $150 billion, a 48 percent increase over 2004. Expanding oil revenues are just what the Saudi government needs to keep a precarious grip on power.

To a large degree, the Saudi royal family, the House of Saud, draws its authority to govern from a commitment to Islamic law and by acting as faithful custodians of the holy cities of Mecca and Medina. Yet over the last several decades the royal family has grown fabulously wealthy by eagerly cooperating with American oil interests. Many from the House of Saud have been educated in the West and spend large parts of their time outside the repressive Islamic theocracy that is Saudi Arabia. This involvement with "all things Western" is perceived as a movement away from the true faith and undermines the legitimacy of the Saudi royal family.

To silence internal criticism, the royal family has struck a Faustian bargain with the Wahhabi Sunni fundamentalists, a xenophobic, militant, and puritanical Islamic sect. As long as its clergy refrains from criticizing the government of Saudi Arabia, the sect can count on lavish subsidies from the House of Saud. Beyond this small restriction, no limits are placed on what can be done to train Wahhabi students or to spread the gospel abroad (Wahhabi missionaries carry diplomatic passports that give them immunity). Flush with oil money funneled through the Saudi government, Wahhabi mosques, madrassas, and Islamic community centers have been built across the globe. On the island of Mindanao in the Philippines are some 3,000 Saudi-funded madrassas. Before 1992, the Taliban regime received Saudi funds. Tens of thousands of madrassas are spread throughout Indonesia, Bangladesh, Pakistan, the Middle East, Morocco, sub-Saharan Africa, and North and South America.

If not for the Wahhabi message, the building of schools, mosques, and community centers would give no reason for concern, but as the Council on Foreign Relations reported in 2004, "Saudi Arabia continues massive spending on fundamentalist religious schools, which export radical extremism that can lead to terrorism." Examples were provided in a May 21, 2006, *Washington Post* article by Nina Shea, drawn from an eighth-grade Saudi text:

> The apes are Jews, the people of the Sabbath; while the swine are the Christians, the infidels of the communion of Jesus.

And a classroom activity is suggested:

> The student writes a composition on the danger of imitating the infidels.

It comes as little surprise that Osama bin Laden and fifteen of the 9/11 hijackers were Saudis.

There is a vicious cycle here, with Western oil money supporting Saudi-sponsored antipathy toward the West. The oil-dependent countries are unable to break the cycle—we need the oil! And the Saudi government is fearful that breaking the cycle will result in their overthrow—a consequence that would have terrible repercussions for oil-consuming nations. It is unthinkable that a fanatical Wahhabi theocracy would be in the position to dictate who should receive oil from one of the world's largest proven reserves, but this is exactly the situation that already exists in Iran.

Iran holds 10 percent of the world's proven oil reserves and the world's second-largest reserve of natural gas. Through Iran, petrodollars from oil and gas exports are diverted to some of the most radical Islamic movements, such as the Lebanese Hezbollah. More alarming, Iran's mullahs seem prepared to use the "oil weapon." In 2002, Ayatollah Ali Khamenei, Iran's supreme leader, threatened, "If the West did not receive oil, their factories would grind to a halt. This will shake the world!" As petroleum demands surge, Iran shows an increasing willingness to leverage its oil, ignoring the United States and its allies in their efforts to prevent Tehran from developing nuclear weapons and sponsoring aggression against Israel.

Winston Churchill cautioned us about the influence oil producers could exercise over a dependent consumer. Nearly one hundred years after this warning, the United States and its allies are now placed in a position of doing business with religious theocracies that deny to their citizens basic human rights, all for our daily oil fix.

Let's turn now to that other thunderhead looming on the horizon, global warming. This black cloud is fundamentally different from that of energy dependency; though global warming is intangible, we can "see" the threat posed through our need of other country's resources. Global warming is something that exists, by and large, in the minds of scientists, and these guys can't seem to get their stories straight. Just thirty years ago, the same scientists who now espouse global warming warned of a coming ice age. It is this public perception of science, climatic science in particular, that enables politicians to wave their hands and say something like "The science is not yet in," and "We cannot pursue a policy based on incomplete information." So, let's look at the science that *is* in.

Carbon dioxide is a greenhouse gas. And there is no question that atmospheric greenhouse gases trap heat; as we have seen, Venus is an incredibly hot, dead planet because of this effect. There is also no question that we are pumping carbon dioxide into the atmosphere. When burning oil, coal, or wood for energy, CO_2 and water vapor result. In 2005, seven billion tons of CO_2 poured forth from tailpipes, smokestacks, and vents. That's more than the combined weight of all the people on earth. In fact, if we loaded everyone on the planet into SUVs—three per car, please—and then piled the whole works onto a scale, we would come close to that seven-billion-ton mark. The question of global warming does not turn on whether or not we are producing CO_2, but where it goes after being produced.

Mechanisms are at work to remove CO_2 from the atmosphere, namely, the carbon cycle (chapter 8). In the absence of evidence to the contrary, it is as reasonable to suspect that this cycle will purge carbon dioxide from the air as it is to suspect the opposite. Quite simply, there are grounds for scientific debate. This was the situation when *Time* magazine ran a June 24, 1974, cover story, "Another Ice Age?":

> Telltale signs [of global cooling] are everywhere—from the unexpected persistence and thickness of pack ice in the waters around Iceland to the southward migration of a warmth-loving creature like the armadillo from the Midwest. Since the 1940s the mean global temperature has dropped about 2.7°F.

The article went on to speculate:

> Man, too, may be somewhat responsible for the cooling trend. The University of Wisconsin's Reid A. Bryson and other climatologists suggest that dust and other particles released into the atmosphere as a result of farming and fuel burning may be blocking more and more sunlight from reaching and heating the surface of the earth.

The *Time* article is often cited to reinforce the idea that "scientists just don't know." Yet somehow the most important paragraph is never referenced:

> But all agree that vastly more information is needed about the major influences on the earth's climate. Indeed, it is to gain such knowledge that 38 ships and 13 aircraft, carrying scientists from almost 70 nations, are now assembling in the Atlantic and elsewhere for a massive 100-day study of the effects of the tropical seas and atmosphere on worldwide weather. The study itself is

only part of an international scientific effort known acronymically as GARP (for Global Atmospheric Research Program).

The scientists of the 1970s recognized that there was not enough information to assess how humankind was affecting climate. Indeed, it was unclear if we were even players. Earth had been marked by dramatic climatic variations before there were people, so why suspect people had some role to play now? Nonetheless, scientists wanting more information became the driving force behind international programs to gather climate information. And one of these programs involved collecting ice cores from the Antarctic.

Antarctic ice is a window into the past. For more than 700,000 years, winter snows have accumulated on this cold southern continent. Summer sun is too weak to melt the snow and so it builds up, year after year, layer after layer, meter upon meter, kilometer upon kilometer. Just like tree rings, each layer is clearly distinguishable from the one formed the previous year. Earth's climatic history can be read in these layers. But it is not their width that does the telling, it is the little bubbles trapped within the ice—atmospheric samples from the year the snow fell. We can analyze these bubbles to determine how much CO_2 was present last year, the year before that, all the way back to a time when our ancestors were chasing antelope across the African veld.

Another piece of data locked in the ice is the average temperature of the world's oceans. To crack this information loose, we need to measure the relative ratios of deuterium (an isotope of hydrogen having one neutron and one proton) to normal hydrogen (no neutrons). Water containing deuterium evaporates at different rates from that containing hydrogen, and the rate difference depends on temperature. By knowing the ratio of these two hydrogen isotopes in Antarctic ice, we can determine the temperature of the ocean from which the water evaporated before falling as winter snow.

The most extensive program to measure ancient atmospheric CO_2

and temperatures is the European Project for Ice Coring in Antarctica (EPICA). After almost ten years of drilling, in December of 2004 researchers pulled the last piece of a 3,190-meter core (about 2 miles long) from the Antarctic ice sheet. In its entirety, this core provides a 720,000-year record of Earth's climate.

Three conclusions can be drawn from this study. First, measurements are remarkably consistent with other ice-core data, gathered at different locations and extracted with different methods. (Such reproducibility provides powerful assurance that what is being measured in these experiments is real.) Second, it shows a clear correlation between atmospheric carbon dioxide and global temperature exists. When CO_2 concentrations go up, so do temperatures. And third, global CO_2 concentrations are now at a 720,000-year high, almost 380 parts per million (ppm).

As a scientist, I find this to be convincing evidence that CO_2 is accumulating in the atmosphere and *will* cause global warming. But I still have doubts; I remember that *Time* cover. To be fully convinced, I would like to see an explanation that accounts for the pre-1974 observations and the ice-core data all at once.

But something else is going on here, which is known as solar dimming. Gerry Stanhill, an English scientist working in Israel, first spotted this effect. Comparing Israeli sunlight records from the 1950s with current ones, Dr. Stanhill found a 22 percent drop in sunlight reaching the ground. Yet this wasn't happening just in Israel; there was a 10 percent drop in the United States, nearly 30 percent in parts of the former Soviet Union, and as much as 16 percent in parts of the British Isles. Overall, there was a 1 to 2 percent global reduction in solar energy reaching the ground every decade between the 1950s and the 1990s.

Dr. Stanhill's observations were greeted with skepticism until Australian scientists, using a completely different method, provided confirmation. The cause of solar dimming appears to be pollution from tiny airborne particles of soot, ash, sulfur compounds, and just plain dust. These visible particles reflect solar radiation back into space, but also change the structure of clouds to make them more reflective. This is just what Dr. Bryson proposed as the mechanism for global cooling back in 1974. Solar dimming is offsetting the effects of global warming. Nothing should surprise us about this, because seldom does a single influence act on a complex phenomenon. Particulate pollution contributes to solar dimming, and CO_2 emissions cause global warming.

Okay, let's pop back to that period before 1974. In 1970, the U.S. Congress passed the Federal Clean Air Act and President Richard Nixon signed it into law. Within just a few years of its signing, particulate emissions by U.S. industries were all but eliminated. Sulfur emissions from coal-burning factories were dramatically reduced. The evidence is clearly seen in the Antarctic ice cores: Post-1970 ice is noticeably cleaner than what came before.

For this skeptical scientist, everything falls into place. The ice age warnings of the 1970s were real, and portended the effects of particulate pollution on global climate. Particulates were cleaned up, but what followed was an acceleration in the rate at which CO_2 was being dumped into the atmosphere. Global warming essentially outpaced solar dimming.

Today, solar dimming is exacerbated by Chinese industries that churn out particulates. The pollution in cities like Beijing, Hong Kong, and Shanghai is much higher than in the industrial cities of the developed world. The offender is exhaust from coal burning, which is passed directly to the air without first cleaning it of particulates, particularly sulfur compounds. The health costs from this kind of pollutant are serious. The World Health Organization (WHO) estimates that particulate pollution accounts for 800,000 premature deaths each year, including 500,000 in Asia alone. So serious is this problem that in 1998 China enacted legislation to curb particulate emissions. Just as

we saw after the U.S. Clean Air Act was passed, China should see a reduction of the particulates responsible for solar dimming and a con-
comitant increase in the effects of carbon dioxide. Every indication is
that the effect will be even more dramatic than the change from global
cooling to warming that characterized the last thirty years.

What might we expect from a world growing hotter as greenhouse
gases accumulate in its atmosphere? We should expect weather to be
more extreme: long dry spells punctuated by torrential rains, violent
tropical storms, and unusual numbers of record-breaking weather
events. Farm production, which relies on stable and predictable weather
patterns, will diminish. Unable to adapt to local climate changes, spe-
cialized and immobile species will vanish. Coral reefs, tropical plants,
and arctic tundra will be among the first to go. Plants and animals more
suited to the changing conditions will expand to fill now-abandoned
ecological niches.

Over the last decade or so, we have observed many of these effects. Does this mean global warming is upon us? Not necessarily. It's
like flipping a coin. It is very rare to see ten heads come up in a row. If
you saw such a thing, you might be tempted to argue that the coin was
not fair—maybe it is heads on both sides—and while this could be
true, there is still the chance that it just happened to come up heads ten
times. As you continue to play, and the coin continues to come up
heads, it becomes more likely that the fix is in, but you will never know
for sure. However, at some point, only a fool will bet on tails. It is the
same with global warming. How long are you willing to play the game
before you conclude the outcome has already been determined?

17

NOW FOR THE GOOD NEWS

C limate change, loss of habitat, oil and gas shortages, declining agricultural production, Western policy response to the dictates of theocratic zealots, uncertainty . . . it's not a rosy forecast. Our problems seem insurmountable, with no clear indication how to proceed. Some think politicians should solve these problems, others that the responsibility belongs to scientists. And still others look to industry for the solutions. I must confess that at times I feel equally confused. Then I remember the *Pioneer* and *Voyager* spacecrafts, and the first step in the development of The Thinking Man's Energy Diet becomes obvious.

Pioneer 10 and *11* were launched in the early 1970s. These space-

craft were the first to leave our solar system to drift for eons in interstellar space. With the possibility that an extraterrestrial might find
one and wonder at its purpose and origin, Carl Sagan and Frank
Drake designed and affixed to each a gold-plated plaque. Etched on
the plaques was the necessary information to locate the planet of the
craft's builders, along with line drawings to represent the builders'
species—a naked couple standing in front of a silhouette of one of the
Pioneer probes. In effect, the plaques serve as an invitation: Here's
where we are, come visit, dress optional.

The invitations became more sophisticated several years later
when, like their predecessors, *Voyager 1* and *2* were prepared for jour-
neys to become interstellar wanderers. To each of these a gold record
was attached, engraved with directions to Earth, but in place of the
naked couple were instructions for playing the record, which is encoded with 115 images and a variety of sounds. Included among these
are the sounds of the surf, wind and thunder, birds, whales, and other
animals, musical selections from different cultures and eras, and spo-
ken greetings from Earthlings in fifty-five languages, along with printed
messages from President Jimmy Carter and UN secretary general Kurt
Waldheim. Many of the 115 images are simple line drawings, convey-
ing such information as the structure of the elements essential to life
and how these are assembled to make DNA. There are also images of
daily human activities—a nursing mother, an athletic competition, and
shoppers at a supermarket.

I remember sitting before my television, enthralled with the pictures from Jupiter, Saturn, Uranus, and Neptune that these amazing
little probes were sending back to a people who wanted to know *more*.
As I watched, I wondered if anyone would accept the invitation to
come visit. Would I turn away from the pictures of Saturn's rings
on my television for the chance to meet a new neighbor? Realisti-
cally, no. So why would other beings be interested in the makers of a
crude spacecraft from a planet located light-years away? In the whole

universe, would the builders of artificial things like *Pioneer* and *Voyager* prove to be so uncommon that space-faring beings would drop everything to find their makers?

On our modest planet, we are not the only builders; in truth, builders are common. In fact, the most prolific of them aren't even people, they are the coral polyps that have fashioned the globe's reefs. The Great Barrier Reef, off the coast of Australia, will become visible to approaching extraterrestrial guests long before the puny constructions of mankind come into view. And though it might be argued that reefs are a physiological product of coral, the constructions of bowerbirds, beavers, and termites are not. What's more, given the option of meeting builders of the cookie-cutter developments that characterize our suburban landscape or the builders of the wonderfully sculptured termite mounds of Africa, might not interstellar travelers choose the latter?

Termite mounds are amazing pieces of engineering. Towering as high as 8 meters (25 feet) into the air, these structures are not where termites live. The mound is instead an air-handling system, serving immense respiratory needs. Just a meter or two below ground, a million worker termites and symbionts collectively consume oxygen at about the same rate as a horse. The mound is designed to capture wind energy, which actively ventilates the nest and maintains its atmosphere. So sophisticated is this ventilation system that only in the last few years have scientists begun to fully understand its subtleties.

Given that termites have acquired a complex ventilation and air-handling system over millions of years of evolution, it is at least plausible that with sufficient time and appropriate selective pressures, termites could fabricate interstellar spacecraft. If so, then the *Pioneer* and *Voyager* space probes might barely raise the interest of our intended guests as they dodge the constructs of numerous space-faring species.

Regardless of how advanced termite technology becomes, there is reason to believe that a desirable race of extraterrestrials will see something in *Pioneer* or *Voyager* that is worth further exploration. It is the invitation, not the invention, that will attract their attention; for in the invitation, in the plaque, is the expression of a species that can ask the question *What if?* What if there are other creatures somewhere in the universe like us? The invitations on *Pioneer* and *Voyager* mark us, each and every one of us, as creatures with an imagination. We are not the only builders on this planet, but we are apparently the only *imaginative* builders. We are the only species capable of building something as a way of answering *What if?* Though termites may one day fashion an advanced ceramic spacecraft propelled by the combustion of methane produced as they digest cellulose, they will not do so as a response to "What if there are planets out there covered with wood?" But it was in pursuit of answers to questions exactly like this that *Pioneer* and *Voyager* were built, and this is why another race of imaginative builders may accept our invitation.

Imagining is what makes us unique among animals. It is not our big brains, per se, or our opposable thumbs, but rather our ability to explore the consequences of the what-if questions. And in the case of energy use, we have already begun to ask them. What if too much carbon dioxide changes the world's climate? What if we run out of oil? What if we have another nuclear accident like Chernobyl?

It is in questions like these that we see what-if's dark side. While visions of a bright future can propel us forward, unintended consequences like pollution, shortages, and dependency can cause us to question, delay, or even retreat. As for our current situation, we can imagine both the means to extricate ourselves from the energy mess, as well as ways to work ourselves into an ever deeper hole. In the end, the picture of a glorious future does not outweigh our fears of potential negative consequences. And, paralyzed by those fears, we do little or nothing about them. Inertia dominates. The present is unwinding along the path set eighty years ago when we saw a future with a car in every garage and a chicken in every pot. Now that those cars may alter our climate, the future is less attractive. Perhaps we're trapped by the same fear that drove the Maya to oblivion.

It was not always this way. Indeed, when I was growing up, imagination shaped the world. President Kennedy's bold vision of landing men on the Moon and returning them initiated a space race that brought us *Pioneer* and *Voyager*. The invitations attached to these craft do not contain images of a sweltering planet or one contaminated with nuclear waste. No, these craft are interstellar messengers carrying our optimistic visions of the present and plans for the future.

This is the lesson I take from *Pioneer* and *Voyager*: If we are to deal with the energy enigma, we must regain a sense of confidence in our abilities to imagine and direct the future, as opposed to being pushed around by the unintended consequences of the technologies we have imagined. Perhaps the first step toward regaining this confidence is to better understand the causes of pessimism.

In the years just after World War II, it seemed that technology could conquer any problem. The Manhattan Project ended the war. The transistor held the potential to revolutionize computers and consumer electronics. A national highway system linking the East and West Coasts was under construction. A vaccine for polio had been discovered, freeing parents and their children from the fear of this paralyzing disease. Effective public health practices were eliminating the scourge of smallpox. Insecticides were stemming malaria, and better fertilizers and herbicides were increasing agricultural yields as more people relocated to urban centers. Coal, oil, nuclear, and hydroelectric energy provided the power to drive a thriving manufacturing sector. Nuclear-powered submarines sailed beneath the ice cap of the Arctic. Chuck Yeager broke the sound barrier. And passenger jets provided rapid transportation from New York to Los Angeles.

Exactly when the optimism that fueled this remarkable period of technological growth began to crumble is unclear, but we can identify significant milestones along the way. In 1962, Rachel Carson's *Silent Spring* hit the shelves and ascended to the top of the *New York Times* best-seller list. The book claimed pesticides were harming the environment, particularly birds, whose eggs were being rendered thin and brittle by DDT. Carson believed that in order to boost profits, the chemical industry was spreading disinformation and misleading public officials. Proposing a biotic approach to agricultural pesticides, she is widely credited with starting the environmental movement that would eventually secure a ban on the use of DDT.[7]

The expansion of the environmental movement was nothing short of astonishing. As public anxiety over what it saw as the degradation of the land, air, and water grew, the science of ecology became the magnifying glass through which to view the impact of technology on nature. Many did not like what they saw, and on November 30, 1969, the *New York Times* carried a lengthy article by Gladwin Hill reporting on the proliferation of environmental events planned to coincide with the first Earth Day.

> Rising concern about the environmental crisis is sweeping the nation's campuses with an intensity that may be on its way to eclipsing student discontent over the war in Vietnam. A national day of observance of environmental problems . . . is being planned for next spring . . . when a nationwide environmental "teach-in" coordinated from the office of Senator Gaylord Nelson is planned.

On April 22, 1970, 20 million demonstrators from thousands of schools and communities turned out in support of the new movement.

As with any good cause, it helps to have villains as a means of focusing and directing outrage. For many environmentalists, chemical and power industries were the scoundrels responsible for the destruction. It had become fashionable, even praiseworthy, to turn our collective imaginations to the negative consequences these industries might mastermind.

Novels and movies about the unintended consequences of out-of-control technology were readily dished up for public consumption. Perhaps the most prophetic of these was *The China Syndrome*—the nickname given to the meltdown of a nuclear reactor's uranium or plutonium core. Released on March 16, 1979, by Columbia and starring Jack Lemmon, Jane Fonda, and Michael Douglas, the film is a political thriller about a news crew that accidentally witnesses a control room crisis at the fictional Ventana nuclear energy plant. In its first two weeks, the movie grossed $20 million (against production costs of roughly $6 million). Just eleven days after its release, on March 28, the reactor core at the Three Mile Island nuclear power plant suffered a partial meltdown, causing America's most serious reactor accident. Jane Fonda began a national tour promoting the movie and crusading against the nuclear industry. Columbia's stock gained five points.

At the same time audiences were flocking to *The China Syndrome*, the residents of a neighborhood in Niagara Falls, New York, were searching for new places to live, having been forced from their modest homes by what was arguably the most appalling man-made environmental tragedy in American history.

In the late 1950s, roughly one hundred homes and a school were built adjacent to a never-completed canal. The 1890s inspiration of William T. Love, the canal was to provide hydroelectric power for a model city situated on the Niagara River. The advent of AC current and an uncooperative economy doomed Love's vision, leaving only a ditch in its place. Sometime in the early 1940s, the Hooker Chemical Company began using this ditch as a landfill, and in 1953, it was covered and sold to the city for one dollar.

For nearly twenty years, the residents of Love Canal were oblivious to the deadly soup percolating beneath their homes. But following a record amount of rainfall one summer, the truth became all too apparent. On the first day of August 1978, the lead paragraph of a front-page story in the *New York Times* read:

> NIAGARA FALLS, N.Y.—Twenty-five years after the Hooker Chemical Company stopped using the Love Canal here as an industrial dump, 82 different compounds, 11 of them suspected carcinogens, have been percolating upward through the soil, their drum containers rotting and leaching their contents into the backyards and basements of 100 homes and a public school built on the banks of the canal.

Eckardt C. Beck, who worked for the Environmental Protection Agency and visited the site, wrote:

> Corroding waste-disposal drums could be seen breaking up through the grounds of backyards. Trees and gardens were turning black and dying. One entire swimming pool had been popped up from its foundation, afloat now on a small sea of chemicals. Puddles of noxious substances were pointed out to me by the residents. Some of these puddles were in their yards, some were in their basements, others yet were on the school grounds. Everywhere the air had a faint, choking smell. Children returned from play with burns on their hands and faces.

Within a week, New York governor Hugh Carey announced that the state would purchase the homes affected by chemicals and, in an action without precedent, President Carter approved emergency financial aid for the residents of Love Canal.

Love Canal was not unique. Many places where industry and government had disposed of waste were resurfacing to exact a tremendous price. "Environmental remediation" and "superfund sites" became part of the modern vernacular. Cleaning these sites would prove expensive, but doing so was not beyond our technological capabilities. A series of tragedies, however, was about to make the public even more wary of science and its technology:

The eighteen months from December 1984 to May 1986 spanned what may be technology's darkest hours. In the early morning of December 3, 1984, a holding tank operated by Union Carbide that contained the toxic chemical methyl isocyanate overheated. The liquid boiled, reacted, and released 40 tons of deadly gases into the heart of Bhopal, India. Heavier than air, these gases rolled through the surrounding neighborhoods, killing thousands outright. Many more were injured, trampled as people fled the lethal cloud. Estimates of those injured range from 150,000 to 600,000, at least 15,000 of whom later died. Little more than a year later, in January 1986, as a television audience of millions watched, the space shuttle *Challenger* exploded on takeoff, killing all seven astronauts on board. Four months after that, on April 26, the Chernobyl disaster hit.

With public sentiment sensitized to technology's undesirable consequences, the *Exxon Valdez* hit a reef in Prince William Sound on March 24, 1989, spilling somewhere between 11 and 30 million gallons of crude oil. Thousands of animals perished. Videos of dead and dying seals, birds, and sea otters were splashed across television screens. By the close of the twentieth century, the technological confidence that had reached its zenith in the years following World War II had collapsed.

As we imagine our energy future against this dismal backdrop, I suggest that we should still turn to science and technology. To many, this sounds like lunacy, since people see technology as the source of the problem. Condensed to its essentials, their argument goes something like this: Were it not for technology, there would be no need for electric power, hence no nuclear reactors, and the horrors of Chernobyl (global warming, air and water pollution, environmental destruction—fill in the blank) would not have occurred. It is impossible to argue against this indisputably true proposition, which leaves the only counter-argument one of comparative advantage. That is, were it not for technology, the people of Chernobyl would be eking out a subsistent existence before dying young (by modern standards) of disease, infection, and malnutrition.

The popular view of technology as "unpredictable" is to be expected from a technologically immature culture. Yes, I believe we are still infants in our understanding of science and technology, and the tragedies associated with their use are those that one would be expected if an unsupervised child plays with things he or she does not understand. I was only three or four when I lit my first match and burned my fingers, and I still marvel that this was the extent of the damage; the house was not destroyed, nor did my clothes catch fire and cause permanent disfigurement. Even though these possibilities are realized numerous times every year, as witnessed on our local news stations, you seldom hear reminders to abandon matches at the close of the story.

The analogy with technological infancy can be made more substantive by considering the development of the stove. Originally intended to provide heat and a means to cook food, the first stoves were open pits where fires were built. Without a means to remove combustion products, these were terribly inefficient and prone to spread fire, as well as posing a serious health hazard by exposing users to carbon monoxide and the carcinogens in smoke. Later, the fire moved from an open pit to a three-sided fireplace. This may have reduced the threat of spreading fire, but it did little to improve efficiency and did nothing to help the problems associated with smoke in living spaces. Chimneys were uncommon in Europe right up to the fifteenth century. It was only in the eighteenth century that stoves built from iron, which fully enclosed the fire, made an appearance. Combustion gases and smoke were expelled via a flue in Benjamin Franklin's famous stove, invented in 1740. The European counterpart, designed by our old friend Count Rumford, came later, in 1798.

Modern fireplaces, stoves, and furnaces suffer from none of their ancestors' deficiencies. Many burn natural gas—as do some fireplaces, which simulate a wood fire—eliminating pollutants. Efficiencies are kept high, because external air is burned with the gas in an airtight combustion chamber. The exhaust is circulated through a heat exchanger and then expelled through double- or triple-walled flues that reduce fire risk. These contemporary appliances are a marvel of technology—safe, efficient, and pollution-free. We grew into these technologies. Isn't it possible that we will grow into many of the technologies developed over the last hundred years—such as chemical manufacture and nuclear power?

In some instances, we have no choice but to grow—and to do so in a hurry. The threat from global climate change is real. Though we do not know if the results will be catastrophic, benign, or maybe beneficial, there *will* be changes. Given the possibilities, we had best not wait to find out what they are. So what do we do? Can we use our amazing imaginations to envision and realize a world where energy is used responsibly?

In the past, we have not been very successful prognosticators. Gas lighting, incandescent bulbs, steamships, gasoline engines, powered flight, electric distribution grids, radio, personal computers, and nuclear energy, to name but a few, were all widely dismissed by the experts of the time. It was fuel requirements—no ship could carry enough—that made transoceanic steamship crossing impossible. The same problem was seen as insurmountable when it came to planes. In 1904, Octave Chanute, an aviation authority, wrote off the possibility that planes would ever carry passengers or cargo over any substantial distance. And in 1900, as cars were embraced in increasingly greater numbers, the historian Edward Byrn believed that man would never be able to "get along without the horse."

At the same time that we failed to see technology's inevitable march forward our willingness to naively accept and promote new energy sources was remarkable. For many nineteenth-century technologists, coal was the perfect fuel from which to reap the benefits afforded by steam engines. The fact that unlimited coal burning would be accompanied by deadly smog, land destruction, mining deaths, and supply limitations was overlooked or ignored. Electricity was the next great boon. Inspired inventors like Edison, Westinghouse, and Ford supplied the world with products that quickly consumed all the hydroelectric power produced. Filling that void was nuclear energy, a supposedly safe and nonpolluting energy source that would make electricity so cheap there would be no need for meters. But the disasters at Three Mile Island and Chernobyl exposed the darkness in fission's sunny future, ending the construction of nuclear power plants across the United States. And then there's fusion. For the last fifty years, its realization has been "just twenty years off." Without harmful waste and from water alone, fusion power will finally satisfy all of humankind's energy needs. Or so they say.

Given our pathetic record, how can anyone argue that technological maturity and the prescience it requires are right around the corner? Perhaps we are expecting the wrong thing. Maturity does not bring with it the ability to look into the future. Rather, it implies experience, which the truly mature individual uses when making future decisions. It is not our ability to predict the course of technological development that will mark our graduation to the next level of sophistication, rather it is how we apply the lessons of the past when imagining and planning for the future. And this is the good news: We have 10,000+ years of experience to apply as we plan.

So, what are these lessons? The first and most obvious is that our technologies form a complex system that demands a constant energy supply. The key here is not found in the energy requirement but in the word *complex*.

A hundred years ago, I doubt there was a single reputable person who would have predicted that using internal combustion engines would cause climate change. The scientists of the time did not have the mindset that would allow them to consider this possibility. Almost everything was thought of as an action followed by a reaction: Push on a weight and it moves; shine the light of the appropriate energy on a metal, and off comes an electron. At the time, science was all about discovering equations that related cause to effect. The very idea that an outcome could not be predicted from a simple set of inputs was anathema to late-nineteenth- and early-twentieth-century science.

The digital computer changed all of this. It was now possible to model phenomena and assess the impact of changing one variable in a system. Simple things like the drips from a faucet or the turbulence from a bee's wings were found not to be simple at all. Apparently there were no "governing equations" to explain these things. Tiny changes to the initial conditions of the system—like the shape of the faucet—produced dramatic and often unpredictable results. Such phenomena are called complex.

The study of complexity, known as chaos theory, became the latest thing with Jeff Goldblum's portrayal of the mathematician Dr. Ian Malcolm in the 1993 blockbuster movie *Jurassic Park*. As a student of chaos, Dr. Malcolm knew that something unforeseeable was going to happen, simply because Jurassic Park was a "complex system." And herein lies some evidence for technological maturity, which I believe we have started to develop. Whereas the scientist of a century ago ignored complexity, scientists of the twenty-first century look for it. Today it would be virtually impossible to bury toxic waste in a Niagara Falls landfill without at least considering future implications. This doesn't mean it couldn't happen, it just means that given the right "technological environment" it would be less likely.

This brings us to our second lesson: Mistakes will happen.

Just as with children and puppies, we should expect mistakes in our technologically immature culture. I have not had the pleasure of raising children, but several dogs have allowed me to raise them. And, I am pleased to say, each canine taught me how to be better with the next. My first dog—her name was Phred—taught me that when it comes to housebreaking, yelling, screaming, and—I'm ashamed to say—hitting are not the way to go. It just makes for a troubled adult dog. The same thing goes for puppies that gnaw furniture, scratch at doors, or growl at mail carriers. One expects these behaviors in puppies. Though punishment alone may "cure" them, they may also become fearful and insecure dogs. Rewarding puppies for being good is just as effective a means of altering behavior, and it makes for happier, healthier adult animals. The same applies to people. Severe punishment of children may alter their behavior, but it also produces adults who are unwilling to admit to mistakes and may even attempt to hide them as a means of avoiding punishment.

What is true for people and our pets is also true of the technology industry: It is immature and will make mistakes. The problem is that our society is intolerant of these mistakes. This is why it took until 1994, after sixteen years of litigation, for the government to secure an agreement from Occidental Chemical (formerly Hooker Chemical) to pay $129 million to help cover the costs of the Love Canal cleanup. And it wasn't one of those friendly agreements where everybody shakes hands and says, "Yeah, we made this mess, we'd better clean it up." Occidental denied liability for years, finally buckling under the pressure of a lawsuit brought by the U.S. Department of Justice and the EPA. But who in their right mind would expect Occidental to admit liability? Other corporations have been bankrupted for far less.

Keep in mind that Hooker started dumping toxic waste into the canal in 1942, right in the middle of World War II. At this time, very little was known about chemical toxicity, the lifetime of these chemicals, or how they moved through the environment. Thirty years later, after Carson's *Silent Spring* and the first Earth Day, the saga of Love Canal began to unfold. Isn't suing, then, a little like threatening to fire an adult from her job after discovering as a child she had scribbled with crayon on her bedroom walls?

I am not suggesting that Occidental should not bear some responsibility for the cleanup. I am wondering why, between 1953 when the landfill was capped and given to the city and 1978 when the immensity of the nightmare became clear, no one came forward to suggest there might be a problem beneath Love Canal. Surely, someone at Occidental knew of the dumping, had read Carson's book, and put two and two together. The reason is simple—our society doesn't reward good behavior. Imagine what a difference it would make if the public were to react to such revelations with "Wow, thanks for bringing this to our attention, Occidental. We had better look into this ASAP." Instead, the response is "How could Occidental have been so negligent and endanger the lives of so many? They'll pay for this one!" We are in conflict with the very technologies that have made us flourish. This is like being at odds with your children or pets—it doesn't make for a happy family.

This brings us to the third lesson: Expect the worst.

Scientists don't know a fraction of what we would like you to believe we know. I think the illusion of scientists as all-knowing began with Albert Einstein. Though he did not have an official publicist, Sir Arthur Eddington did an outstanding job of propelling Einstein into the limelight. In his 1915 Theory of General Relativity, Einstein predicted that light was just like everything else and would "fall" in a gravitational field. Because it moves so fast, light doesn't fall straight toward most objects; its path is just bent a little. Eddington realized that measuring the deflection of starlight passing the sun could test this prediction. But only during a total eclipse would it be possible to make such a measurement, because at all other times the dim stars would be obscured by the much brighter sun. In 1919, Eddington headed an expedition to do just that—measure the deviation of starlight as it passed near the edge of the eclipsed sun. The results confirmed Einstein's predictions.

The press's reaction to Eddington's announcement was extraordinary. Einstein was the lead story of the world's media and, almost overnight, he became a household name. Perhaps after years of war and destruction, the public was ready to embrace an act of pure mind and imagination that lay bare the secrets of the cosmos. For whatever reason, scientists were now larger than life, capable of any feat.

The notion of the all-knowing scientist and engineer was perpetuated within movies, literature, and television. How many times did Captain Kirk beam back to the *Enterprise* with an appropriated piece of alien technology and hand it to Scotty? With only a look and a wink, Scotty would say something like "Aye, Captain, we'll have to wire this directly into the warp core. Give me two minutes," and in less than two minutes, it was done. The starship *Enterprise* and her crew were saved yet again. I don't know about you, but it takes me more than two minutes to screw in a couple of lightbulbs. If I had to adapt an alien lightbulb to work in an earth lamp, you had better give me a week or two.

The truth is that science allows us to narrow the realm of what should be expected, but not to predict the future. I believe that greenhouse gases are trapping more heat in Earth's atmosphere. The data are overwhelming. Some climate models predict that this heat will lead to desertification; others do not. The disagreements have led many to ignore the whole thing. Basically, they are saying, "When the scientists get their acts together, then I will worry." You see, science is "supposed" to be certain, so as long as there is uncertainty, it can't be science. This is wrong, wrong, wrong!

Don't seize on scientific differences to make yourself complacent. Instead, look for the agreed-upon facts. What is agreed upon is that all the factors that could cause climate change are present. If I were told to prepare for a trip to an unknown destination, I would pack for the extremes: shorts and sunscreen, umbrella and waterproof boots, and a parka and gloves. I would prepare for the worst.

So, The Thinking Man's Energy Diet will be built as a complex system of energy-generating technology. The builders of this system will be charged with anticipating and eliminating errors, regardless of the cost—because even small errors may have disastrous consequences. Even so, we expect that there will be problems, and as mature people, we have come to know that it is better to reward those who discover problems than to punish those who caused them.

One more small detail remains. What will this system look like? How will it function? For that, we must rely on imagination. But whose? I suggest that we all play a part in imagining our energy future. In the past, we have left it to the government. Unfortunately, too often those who govern confuse imagining with wishing.

Our Energy Future

18

THE THINKING MAN'S
ENERGY DIET

There is a bright line separating imagining from wishing. Anyone can wish for a renewable and secure energy future, but without a plan to get there, it remains only a wish. Sure, wishes come true occasionally. For a few days, or weeks, in 1989, all of our hopes for abundant and clean energy seemed to have been realized: Martin Fleischmann and Stanley Pons, chemists from the University of Utah, announced they had induced fusion with a simple electrochemical cell using heavy water (water where the hydrogen is replaced by its deuterium isotope) and a palladium electrode. Dubbed *cold fusion,* Pons and Fleishmann claimed to have initiated a process at room temperature that most physicists believed would begin only above tens of millions

Kelvin. If the chemists were right, the abundant energy source that powers the Sun and stars would be readily available to feed the fires of our imaginations.

Regrettably, the announcement was premature. Most scientists have come to believe that cold fusion is simply wishful thinking, though there are those who continue to explore its potential.

The cold fusion episode and our reaction to it provide an excellent example of what is, effectively, the energy policy of the United States. In the back of our leaders' minds the belief dwells that if the inevitable can be held off long enough, science and technology will come to the rescue. In other words, they are wishing for a solution. As the cost of energy climbs ever higher, they firmly believe the innovators of the free market will ride in on white horses. They wish oil companies will find new reserves. They wish bright young scientists working at the underfunded National Renewable Energy Laboratory (NREL) in Golden, Colorado, will discover solar cells that are so efficient and cheap that they will eliminate our dependence on oil from the Middle East. They wish some clever modeler at the underfunded National Center for Atmospheric Research (NCAR) in Boulder will discover that global warming is a natural process caused by the juxtaposition of sunspots and the regular variations of Earth's orbit. They wish investigators at an industrial research lab will figure out a way to mimic bacteria that use sunlight to make hydrogen, while another group of scientists at the Ford Research Center design revolutionary hydrogen fuel cells to power our cars.

The problem with this wishful approach to energy "policy" is that wishes seldom come true. I am sure the Maya were praying to their gods and wishing for rain even as their crops died and the forest continued to retreat. The people of medieval England probably wished that coal wasn't a nasty, smelly, choking alternative to wood and charcoal. And we all wish that a gallon of gas sold for what it did ten years ago.

Imagining is quite different. It involves planning. Had the people of London imagined coal as a replacement for wood, they would have explored ways to mitigate its smelly smoke. They would have discovered coke, and town gas. Instead, they stumbled upon these energy alternatives after becoming engulfed in killer smog.

You might think that in today's technologically advanced world, we plan for our energy future. We anticipate problems related to production and utilization, and we imagine and test strategies to moderate these effects. Well, let's see.

Developing the 800 billion barrels of "crude" suspended in the oil shale of the Green River Basin straddling Colorado, Utah, and Wyoming is confounded by a number of costly technological issues. Consequently, only when oil prices exceed some minimum does oil shale become a profitable resource. This was the case in the late 1970s when oil from Iran was embargoed and crude prices hit $40 per barrel with a $70 forecast. At these prices, it was profitable for Exxon to strip-mine tons of oil shale from the Green River Basin and, through a costly and potentially environment-threatening refining process, extract the oil.

In anticipation of the new industry, land values in the rural communities around the Green River escalated almost beyond reason. Ranchers sold their land for ten times its pre–oil shale boom value of $200 an acre. The Colorado towns of Parachute, Rifle, and Silt were struggling to keep up with the projected population explosion by building schools, paving roads, and extending utilities. At the peak of the speculation, the population of Rifle was growing by 50 percent a year. The oil industry was heading for Colorado, and Denver was experiencing an unprecedented building surge, with construction cranes all but obliterating views of the Rocky Mountains from parts of the capital city.

As the Iran-Iraq war geared up, and both sides flooded the oil markets to pay for their aggression, the price of crude went south, plummeting to below $20 per barrel. On Black Sunday (May 2, 1982), Exxon dropped its massive Colony oil shale project, abandoning $5 billion in development costs and laying off 2,200 workers. Oil companies fled Denver for Houston, leaving new skyscrapers barren. Within months, office space in Denver was the least expensive in the world, with the possible exception of Kuala Lumpur, Malaysia. Ten years later, Colorado and Denver were just beginning to recover from the trauma of Black Sunday.

In 2005, the price of crude oil was once again bumping against that ceiling in which shale becomes a "good investment," and the oil companies were back. What's telling is that most analysts foresaw their return. Even as the oil companies packed their bags and fled from billions in investments, it was clear: Oil shale was a fuel source of the future, and the technological and environmental problems associated with its extraction and distribution would have to be solved. When the oil companies returned, they started rebuilding their shale programs by seeking out the brainwork of those who had studied those problems thirty years earlier. What they found was that many had retired, some had died, and much of this work had been forgotten—or worse yet, lost.

While it is sensible to base corporate strategy on economic viability, it is unacceptable that national energy policy should be held hostage to profits. Developing the 800 billion barrels of crude suspended in the Green River Basin would provide a means to offset reductions in world supply brought about by embargos, military attacks on pipelines and other sources of supply, or natural disasters, such as earthquakes or hurricanes. In short, it would be a means to stabilize crude's volatility and counter the oil weapon.

If Washington had been "imagining" an energy future independent of foreign oil, wouldn't it have made sense to exploit the brainwork of scientists and engineers who had devoted years to solving the problems associated with converting oil shale into oil? The years between Black Sunday and today could have been used profitably and productively. If the Soviets had embarked on a weapons program designed to isolate the United States from energy resources, we would have responded. Because when it comes to National Security, all facets of society—industry, government, and individuals—work together toward a common goal.

Here is one of the puzzles surrounding America's energy strategy: Our policymakers tell us that energy is an economic resource, and they leave it to the oil, coal, gas, and electric suppliers to imagine the energy future. At the same time, we divert huge amounts of our defense spending to guarantee access to oil from the Middle East. The truth is that energy independence is the security concern of the twenty-first century, just as the struggle with the Soviet Union was the concern of the latter half of the last century. We should take some measure of satisfaction in this fact, for when it comes to imagining a *politically* secure future, the government and people of the United States are unmatched.

The Cold War provided forty-five years to practice and refine our abilities to imagine. During this time, at least as far as defense technology was concerned, confidence in our ability to shape the future never failed. Government and defense industries seldom fell into the trap of finger-pointing and blaming others for mistakes. The emphasis was always on fixing problems and moving forward. This "imagining-friendly"environment sprang from a national unanimity of purpose. Both major political parties recognized the threat posed by Soviet territorial ambitions. Accordingly, the Department of Defense was unfettered in its development of weapon systems to deter Soviet expansion. All branches of the military embarked on long-range planning, specifying capabilities and a timetable for acquiring them. The evolution of the F-4 Phantom fighter of the 1960s into today's F-22A Raptor proceeded along steps of a carefully orchestrated research and development program. Planning of this sort was possible because defense policy did not shift with each election cycle. In turn, only an electorate that shared a common vision of the future could provide such policy stability.

Some have suggested that something along the lines of an Apollo or Manhattan project should be instituted to extricate the United States from its energy predicament. However, landing a man on the moon and developing an atomic bomb are insignificant tasks compared to realizing a secure and renewable energy future. To liberate ourselves from energy dependency and protect ourselves from the threat of global warming, we must attack the problems with the same commitment and solidarity we displayed during the Cold War.

Nowhere is that solidarity of purpose more typified than in the emergence of the military-industrial complex. This "union" of government and industries was a natural one, intended to drive technologies imagined as critical for national security. Take the nuclear U.S. Navy, for example. The first nuclear-powered ship (a submarine) was the USS *Nautilus,* which put to sea in 1955. Currently, the navy operates more than eighty ship-based nuclear reactors and has accumulated over 5,400 reactor years of *accident-free* experience—the best record on the planet.* This unique record of reliability and safety results from the way the military-industrial complex functions. By understanding its operation, we should be able to build a similar system for The Thinking Man's Energy Diet. So let's look in a little more detail at the relationship between the military and its contractors.

Weapons development, in general, and reactor design, specifically, begins as the military (in this case, the U.S. Navy) joins with manufacturers to develop long-range plans. These plans may specify reactor performance criteria, say, for the year 2035, as an example. The technological hurdles to reach the desired goal are identified, along with target dates for surmounting each hurdle. The navy, through its Office

---

*There are 64 nuclear power plants operating within the United States. France, with the world's most ambitious civilian nuclear program, operates 59.

of Naval Research, takes on the challenges of the distant objectives, leaving it to the manufacturers to tackle the more immediate problems.* At intervals during the development process, the navy calls on manufacturers to submit bids for reactors with improved performance, which will be used to upgrade existing ships or power new ones. The bid is awarded based on product reliability, performance, and cost. It is common to base an award on performance or reliability even when the cost is higher. Manufacturers participate in this relationship because the government is providing them with a market years in advance. All that is required to reap the benefits is to innovate according to the navy's timeline. (Consider the boon to an automobile manufacturer that knows what the public is going to want right up to the 2035 model year.) In return, the navy gets a continually evolving and improving product. What makes the whole thing possible is the navy's ability to set distant objectives.

Let's do exactly the same thing with The Thinking Man's Energy Diet by creating an energy-industrial complex. Start by imagining a future thirty years hence, and specify timetables to achieve specific goals. Make industry a part of the process by helping manufacturers tailor their market for the foreseeable future, and build in financial rewards for those manufacturers that come closest to meeting the goals.

Envisioning America's energy use in 2035 is easy: America will be energy independent. The energy we use, we will produce. Through brainwork, we will be the world's leading exporter of carbon-free energy and of energy-efficient products. We will have made the oil weapon obsolete.

Now comes the hard part—formulating the plan to achieve this secure and renewable energy future. Clearly, a complete plan is beyond the scope of this book, or for that matter any single book. Ultimately,

---

*Some of this research the navy does in-house, and some is contracted out to national laboratories, universities, and corporate research labs.

the brainwork of thousands will be integrated before constructing a national energy policy. Therefore I will outline only the barest of ideas for achieving renewable energy independence.

My approach is to abandon internal combustion engines in favor of electric motors. Recall that Carnot reasoned that heat engines (internal combustion engines) waste energy. As we have seen, this is one statement of the Second Law of Thermodynamics. Accordingly, a heat engine cannot be 100 percent efficient. The engine in most automobiles is between 20 and 25 percent efficient, meaning that the remaining 75 to 80 percent of the input energy is turned to heat before reaching the drivetrain. On the other hand, electric motors are not heat engines and therefore are not subject to Carnot's rules. In principle, an electric motor could be almost 100 percent efficient, though in practice it will fall short of this goal. After all, a running electric motor generates heat.

Efficiency, however, only addresses the Second Law. The First Law tells us that we still need an energy source from which to make the electricity powering the world of 2035. So, our energy diet plan must contain two parts: a process to replace internal combustion engines with electric motors and a strategy to generate electricity cleanly and without reliance on foreign sources of oil. As this two-part plan must proceed in step from today's technologies, let's review some of these— particularly those concerned with electric transportation and electric generation.

As far as electric transportation is concerned, there are essentially three ways to supply power to a vehicle: 1) through a remote generator; 2) through onboard rechargeable batteries; and 3) through onboard generation.

Rapid transit systems and commuter railways generally employ the first of these strategies, directing electrical power to the train via a third rail or overhead lines. Unfortunately, too many people in the United States associate electric rail systems with old trolley cars, like those of Boston's Green Line that I rode as a graduate student in the '70s and '80s. These uncomfortable and slow-moving vehicles give the wrong impression of electric transportation. A much better example is of the modern bullet trains, like the Eurostar that connects London to Paris. The two-hour trip between these cities is best described as peaceful. The train rolls on welded rails supported by steel and concrete ties, giving passengers almost no sensation of motion—even at speeds in excess of 300 kph (186 mph).

Although there is nothing to prevent powering electric cars and trucks with a third rail, batteries are the preferred source of energy. Until recently, the battery of necessity was the common lead-acid rechargeable. This was the battery used in the EV-1, made by General Motors between 1996 and 1999. It produced a top speed of 80 mph and a range of about 70 miles. This worked fine for the business commuter who worked no farther than 35 miles from home. But even then, each night the EV-1's 26 lead-acid batteries—together weighing in at 1,310 pounds—had to undergo a several-hour recharge.

Over the last ten years, the power packed in a rechargeable battery has increased. Now with a 900-pound lithium ion battery pack, the Tesla Roadster—with an equivalent fuel efficiency of 135 mpg—will carry its occupants 250 miles between charges. In case you think "scooter" when your hear 135 mpg, just hit the accelerator of the Tesla and feel yourself slammed against the seat by the four-second, zero-60 acceleration. (That's just a half-second slower, and a whole lot quieter, than the 500 hp 2006 Corvette Z06.)

Hybrid cars use a gas-burning engine along with onboard electricity generation to charge batteries that provide energy for a supplemental electric motor. Hybrids, such as Toyota's Prius, take advantage of regenerative braking, in which the energy that would have gone to heat brake pads and generate entropy instead powers an electric generator. In slow traffic or when idling only the electric motor is used; when required, both the electric motor and internal combustion engine can work in concert to provide the power needed for acceleration or climbing.

A plug-in hybrid can be recharged with household current from an electric grid. These vehicles generally have batteries that are large enough to allow operation in electric-only mode for short commutes and still provide the range afforded by an internal combustion engine.

Finally, sharing the advantages of battery power with quick and convenient recharging, are fuel cell vehicles. Fuel cells, like batteries, derive energy through chemical reactions. In the case of rechargeable batteries, the chemical reaction can be reversed with an electric current, restoring the battery power. For nonrechargeable batteries and fuel cells, the chemical reaction cannot be reversed, but fuel cells *can* be refueled.

Most fuel cells convert the energy of combustion directly into electrical energy. Almost any conventional fuel can be used—ordinary gasoline, methane, propane, and hydrogen are a few examples. Oxygen from the air is made to react with these molecules to produce carbon dioxide and water vapor (when using hydrogen, only water vapor is produced), which is expelled into the environment. When all the fuel is "burned," the cell can be refueled just like a conventional engine. If it were running on gasoline, the attendant at the corner service station could "fill it up."

Fuel cells operate by separating the fuel from the air's oxygen with a thin membrane, of which there are two basic types. The first is called a PEM (proton exchange membrane) and the other a SOM (solid oxide membrane). PEM fuel cells are powered by hydrogen gas, and those using a SOM can also be fueled with gasoline, ethanol, methane, and other hydrocarbons.

Before electric cars and trucks become competitive with their internal combustion counterparts, batteries and fuel cells must be made smaller, lighter, more reliable, and less expensive. And here is the plan to get there:

Standardize—require that the battery packs of hybrids and all-electric cars be interchangeable with one another and with fuel cells. This will allow electric and hybrid owners to upgrade their vehicles simply by replacing the power source. Today it's a fuel cell car, tomorrow an all-electric car; today it's a plug-in hybrid with a range of 20 miles on batteries alone, tomorrow 40 miles.

Next, provide a guaranteed market for battery and fuel cell manufacturers by requiring cars and trucks to achieve minimum energy-efficiency standards. These are not the same as fuel-efficiency standards, like CAFE (Corporate Average Fuel Economy), but simple thermodynamic efficiencies. If you want a car that goes zero to 60 in two seconds, that's fine, but it had better do so efficiently, with a minimum of heat generation by the power source and motor. Set these standards in advance. For example, by 2035 we want 75 percent efficiency; in 2030, 70 percent, etc. At the same time, other targets should be specified, like power-to-weight ratio, recharging times, and so on.

Offer economic incentives to manufacturers that surpass the established standards. For example, provide a 10 percent tax rebate for every percent by which a manufacturer exceeds efficiency standards. As an example, suppose the regulations call for fuel cells to be 50 percent efficent and manufacturer A conforms with this regulation exactly, while manufacturer B produces a 55 percent efficient fuel cell. Manufacturer A sells its product for $500, but manufacturer B can sell its better product for $1,000, since the buyer of B gets a 50 percent tax rebate or $500; hence, as far as the consumer is concerned, the two products cost the same. Car junkies will then anticipate the battery and fuel cell manufacturers' announcement of efficiency specification for the upcoming model year. And families will save for the newest and most powerful "fast-charge battery pack" that will turn the old car into an electric muscle machine.

Finally, we should embark on a strategy to equip major highways with third rails or other means to supply power directly to electric vehicles by 2035. Cars could then get to the interstate on power from batteries or fuel cells, and there latch on to the third rail and cross the country without the need to recharge or refuel.

Where will the electricity to power third rails and recharge cars come from? At present, about 40 percent of the energy consumed in the United States is used to generate electricity. Of this, only one-third finds its way to the end user in the form of electrical power. The rest is lost while converting chemical energy—coal, oil, and natural gas—into electricity. Thus by 2035 there will be a need for more electricity, generated more efficiently with a reduction in the carbon emissions that come from making it.

A way to realize these objectives is by stimulating the growth of renewable energy sources, particularly wind and solar, and at the same time expanding nuclear generating capacity. While nuclear energy is not technically renewable, it is virtually unlimited.*

Wind is a plentiful energy resource. There is enough wind energy in the United States to generate 1,250 gigawatts, and some of the best places to harvest this energy are in the Great Plains and along the ridges of the Rocky Mountains. The bounty of this resource remains almost untapped, with a generating capacity of only 9 GWe in 2005. However, spurred by a 1.9-cent per kilowatt/hour tax subsidy for producers, growth has been dramatic—over 36 percent in the last five years. Unfortunately, as a means of increasing tax revenues, Congress has repeatedly threatened to cancel the tax credit.

Solar energy can be collected in many ways, but in the context of electricity generation, the most promising approach is with solar cells (also known as photovoltaics). These devices capture the energy of sunlight and convert it directly to electricity. One of the attractions of solar cells is that they can be installed in a variety of places, including

---

*Bernard Cohen, physics professor emeritus at the University of Pittsburgh, has argued that nuclear energy from uranium will last for 5 billion years.

the roofs and walls of buildings. California is part of the Million Solar Roofs Initiative that intends to create 3 GWe of solar-generating capacity in the state by 2018. (The current worldwide generating capacity of solar cells is around 5 GWe.) The principal impediment to a continued and accelerating expansion of photovoltaics is expense: Electricity from solar cells now costs about four times that produced in coal-fired power plants.

Our policies are disjointed—an uncertain tax credit here, a state effort there. The U.S. Navy doesn't develop one policy for destroyers and another for cruisers. All navy ships work toward a common goal. We should do the same thing by bringing all parts of the electricity-generating infrastructure under a common roof. Using goal setting and economic rewards, a plan can be devised that stimulates renewable energy technologies. For example, we may conclude that by 2035 a national generating capacity of 3,000 GWe will be needed, and of this solar, wind, and nuclear should respectively provide 20, 25, and 30 percent. Government will then partner with industry to identify inter-mediate benchmarks and product performance objectives. In this way, the manufacturing sector, not individual manufacturers, will be guaranteed a market. By way of illustration, if the U.S. government pledges to purchase, or in some manner subsidize the purchase of, solar cells with a net capacity of 20 GWe by 2010, you can bet solar cell manufac-turers will do their best to get a piece of this contract. The net effect will be to push solar cell technology while driving costs down.

This is exactly how advances in integrated circuit production gave birth to an industry. The first integrated circuits were invented in the late 1950s but were used only when cost was not a factor. When it came to reliable electronics for the control and guidance aboard mis-siles, especially the Minuteman intercontinental ballistic missile of the 1960s, security trumped cost. The U.S. Air Force demanded inte-grated circuits. Electronics industries responded. Manufacturers inno-vated to capture a larger part of military contracts, driving costs down.

Simultaneously, the silicon-processing technologies that enabled the consumer electronics revolution were spawned. Every computer, MP3 player, microwave oven, and television owes its existence to a primitive integrated circuit in a 1960s-era missile.

It is not necessary to look far for a model from which to design a nuclear power program. As mentioned, the U.S. Navy has operated more nuclear reactors than any other entity, with, as far as is known, a clean record. Such a program would differ from the present nuclear power initiative, Nuclear Power 2010, in that focus would be on safety and reliability instead of economic viability.

Nuclear Power 2010 links government and industry in an effort to identify sites for new nuclear power plants and explore advanced nuclear technologies. So far, so good. However, the program also calls for the evaluation of the *business* case for building an advanced nuclear power plant. The evaluation will include a study of the nuclear regulatory process, which in the past has led to delays and escalating costs for those companies bringing new reactors online.

We don't make a business case for aircraft carrier battle groups, and we shouldn't be doing it for nuclear power plants that are as important to national security as naval ships and planes. Why not have the navy (or similar entity) operate our nuclear facilities? At first blush, this may seem an absurd idea, antithetical to the free market system; recall, however, that the safety of American air passengers is entrusted to the Federal Aviation Administration. Air traffic controllers are employees of the U.S. Department of Transportation, and the FAA operates the safest transportation system in the world. Just as the federal government exerts direct control to assure safe and secure air transportation, the same should be true of energy—particularly nuclear energy.

Generating more electricity with solar cells, wind power, and nuclear energy will take time. Moving from internal combustion engines to electric motors will take time. In the short term, the oil weapon still looms; in the interim, there is a solution: coal.

Coal is a paradoxical fuel. In the course of its retrieval, miners die and landscapes are scarred. Its use causes acid rain and pollutes the environment with toxic impurities such as mercury. Per unit of energy produced, it releases more carbon dioxide than any other fossil fuel. However, there is a lot of it. The world has more than a trillion tons of readily available coal. The largest portion of that (25 percent) is found in the United States. You might say we're the Middle East of coal.

With coal abundant and inexpensive, it is the fuel of choice for electricity generation, accounting for roughly half of the electricity generated in the United States. Over the next thirty years, it is estimated that power providers will construct coal-fired power plants with a combined generating capacity in excess of 150 GWe. And that's just in the United States. China is building the equivalent of one large coal-fueled power station each week. If this continues, by 2035 it will be virtually impossible to hold atmospheric CO_2 concentrations below the 450 ppm level that many scientists believe to be "tolerable."

We could stop generating electricity with coal, but then it would be unlikely that we could meet our 2035 goals. The answer is found in technologies that allow carbon dioxide to be isolated from the atmosphere. Called carbon capture and storage (CCS), or geological carbon sequestration, carbon dioxide is separated from the reaction products that result when energy is extracted from coal. The CO_2 is then stored deep underground in porous rocks. Ironically, depleted oil and gas fields are perfect for this purpose. Energy-depleted carbon is returned to the same vaults where nature stored it as energy-rich fossil fuels in the first place.

Separating CO_2 from coal's combustion products is not an easy matter, and as it turns out, unnecessary. The more direct and least expensive way to collect CO_2 is through the integrated gasification combined cycle (IGCC). This cycle begins by combining pulverized coal, water, and oxygen in a high-pressure reactor. These ingredients combine to produce a gas mixture of hydrogen and carbon monoxide (CO) called syngas. The pollutants common to coal—mercury, sulfur—are more easily removed from syngas, which is then combined with hot steam to make hydrogen and CO_2. The clean and carbon-free hydrogen is shuttled on as the combustible fuel powering electric generators. The remaining CO_2 can be pumped to underground storage.

Overall, these plants are right around 60 percent efficient. There is an energy cost associated with the carbon sequestration, but it's a small one. Ten to twenty years in the future, efficiency gains could be achieved by replacing turbine generators with hydrogen fuel cells.

What is stopping the development of IGCC power plants? It is the high capital cost associated with construction. Until these costs come down, or the price of carbon-spewing power plants goes up, a combined IGCC-CCS facility is not economically viable.

Once more, should it be economically viable? There should be policies in place that reward companies for doing the right thing. These same policies should advance technologies related to national security. Clean and reliable energy is a matter of national security!

This brings us to biomass. Thanks to a $2 billion annual subsidy, almost 3 percent of our automotive fuel comes in the form of ethanol, the same molecule responsible for inebriating millions of people annually. Fermenting and distilling corn accounts for the vast majority of ethanol produced. There is some question, however, as to whether ethanol from corn produces or consumes energy.

When agricultural products are used to "feed" internal combustion engines, the Great Energy Rule comes back into play. It takes energy to plant, irrigate, fertilize, harvest, transport, and distill corn. If this is greater than the energy derived by burning the alcohol, the ethanol subsidy doesn't make thermodynamic sense. Estimates vary; some policymakers have asserted that ethanol from corn is a net loser. The most optimistic appraisal suggests there is a net yield of one unit of energy per four invested; that is, production of ethanol from corn is 25 percent efficient. (For comparison, oil derived from oil shale is estimated to be between 40 and 60 percent efficient.) Even if the optimistic estimate proves to be right, is this the best place to invest $2 billion annually?

The technological hurdle confronting biomass is limiting waste. Only a tiny portion of the sun's energy absorbed by a corn plant finds its way into ethanol. The energy in the stock, leaves, and cobs are wasted. Only the kernels are turned into fuel. There are ways to convert the cellulose in the waste into ethanol. Research and development intended to stimulate technologies to produce cellulosic ethanol would go a long way toward making biomass an efficient alternative energy source. Maybe this is where that $2 billon should go.

So far, our energy diet plan has been concerned with how the United States will use energy in 2035. We cannot, however, proceed in a vacuum. For two billion people in developing countries there is no electricity. There are no lights, refrigerators, running water, televisions, or access to the World Wide Web. Unable to switch on a light or stove, millions of people, mostly women and children, spend their days gathering wood or other combustibles for cooking, heating, and illumination. Unable to coax running water from a faucet, they spend hours carrying water from distant lakes and rivers, often contaminated with bacteria. We must supply these people with the energy they need to fuel the fires of their imaginations.

A small panel of solar cells and a battery pack would provide the light to read or to learn to read by at night. Larger arrays could power a whole village and provide the power to purify and pump water. Small and safe nuclear reactors could provide energy for many villages or a small city. Hydrogen, produced in ICCG power plants, could be exported to provide fuel for cars and factories where electricity generation is impractical. In short, the United States should become the energy supplier to the world.

The last part of The Thinking Man's Energy Diet is perhaps the most important: We must fulfill our obligation to Nature and continue to imagine. We should imagine fusion energy and find ways to make it real. We should imagine giant generators driven by the flow of ocean currents, and geothermal power plants that extract energy from the heat kilometers below Earth's surface. We should imagine generators orbiting the Earth that transmit clean energy to waiting cities below. We should imagine using all of this energy efficiently—doing the greatest amount of work possible while generating as little heat as the Second Law of Thermodynamics will allow. We should imagine rail systems that move freight as efficiently as UPS moves parcels and efficient trains that connect major population centers. We should imagine—not wish—that every person on the planet has access to the clean and abundant energy they need to feed the fires of their imaginations.

The vision I have outlined is but one of many possibilities. Just as there were numerous plans that could have taken mankind to the moon, and several that would have brought us to victory in the Cold War, we settled on one and made it real. We must do that again.

To borrow from President Kennedy one more time, I urge that we choose a secure and renewable energy future, not because the choice is easy, but because it is hard, because this goal will serve to organize and measure the best of our energies and skills, because this challenge is one that we are willing to accept, one we are unwilling to postpone, and one that we intend to win.

Notes

1. Logarithms are mathematical operators that essentialy tell you how many digits there are in a number. As an illustration, the logarithm of 10 is 1; the logarithm of 100 is 2; and the logarithm of 1,000 is 3. In symbols, $\log(10) = 1$; $\log(100) = 2$; and $\log(1000) = 3$. Numbers between 10 and 100 have logarithms between 1 and 2. Those between 100 and 1,000 have logarithms between 2 and 3, etc. For example, $\log(50) = 1.699$ and $\log(500) = 2.699$.

   With this definition, logarithms have the following property:

   $$\log(A \times B) = \log(A) + \log(B).$$

   Just to verify the validity of this equation, let's try a simple numerical example:

$$\log(500) = \log(50 \times 10) = \log(50) + \log(10) = 1.699 + 1 = 2.699$$

Exactly what we expected. Also, there are two different kinds of logarithms: natural and base ten, denoted **ln** and **log** respectively. The two forms are related as follows:

$$\ln(A) = 2.303 \log(A)$$

Boltzmann knew that entropy had to have the same properties as logarithms. We can follow his reasoning by thinking about a pair of dice. There are six possible outcomes for the roll of a single die: 1, 2, 3, 4, 5, or 6. That is, there are six *ways* a die can roll "up." However, there are thirty-six outcomes for a pair of dice. For each of the six possibilities for the first die, there are six possibilities for the second, giving 6×6, or 36 total outcomes. We can calculate the entropy for a single rolling die using Boltzmann's formula as follows:

$$S_1 = k \log(6)$$
$$S_2 = k \log(6)$$

Here S_1 and S_2 are the entropies for each die in the pair, and k is a constant. Now, by the same formula, the entropy for the pair is:

$$S_{12} = k \log(36)$$

We want the sum of the entropies of two single die to equal that of the pair, just as the energy of two things equals the sum of their energies (the energy in two gallons of gas is equal to twice the energy in one gallon). In equation form:

$$S_{12} = S_1 + S_2$$

This is true only if entropy is given by logarithms:

$$S_{12} = k \log(36) = k \log(6) + k \log(6) = S_1 + S_2$$

By pure reason, Boltzmann was able to see that if entropy was related to the number of ways a system's energy could be realized, it had to be proportional to the logarithm of that number. For mathematical reasons, he chose to express his famous equation in terms of the natural logarithm. Had he chosen the base 10 logarithms, then the Boltzmann constant would take on a different value.

2. After the universe became transparent, the photons that constituted its "hot glow" expanded along with the rest of the cosmos. As these photons moved farther apart, energy of motion was converted to potential energy, just like an object moving upward in a gravitational field slows down. However, a photon cannot slow down, it has to move at the speed of light (300,000 kps), so to conserve energy, the wavelength of the photons increases. This phenomenon is called *red shift*. Over 14 billion years, the photons that were visible have become microwaves, and collectively are known as the *cosmic microwave background radiation*. The average wavelength of this radiation is about six centimeters, the same wavelength that televisions receive. So, if your television is hooked up to an antenna, you can turn to a channel where there is no broadcast signal and watch the static. You are now looking at an image that originally aired almost 14 billion years ago.

3. One of the questions that confounds modern science is how the universe became lumpy. It is not a smooth uniform distribution of energy and matter. There are stars and galaxies suspended in what is an otherwise empty cosmos. If the universe were uniform, it would all look like the space between galaxies—basically empty. The cosmic microwave background radiation gives us one way to determine when the universe became lumpy. We can survey this radiation to see if it, too, shows variations. If it does, we know the universe was inhomogeneous sometime before it became transparent—a few hundred thousand years after the Big Bang. If it doesn't, then theorists would be hard-pressed to explain the origins of the universe's lumps. This is the task given to the COBE (Cosmic Background Explorer) satellite—to measure the cosmic microwave background radiation in every direction and look for variations. It was launched in November of 1989. As anticipated, the satellite detected small variations in the background radiation on the order of one part in 100,000. Each of these "little" variations evolved into the bigger variations that gave rise to galaxies, stars, and planets.

4. In 1953, Stanley L. Miller and Harold C. Urey of the University of Chicago conducted a now famous experiment to explore the origins of

life. They sealed water, methane, ammonia, and hydrogen (all thought to be present in Earth's early atmosphere) in a sterile flask. Sparks were shot through the mixture to simulate lightning. At the end of one week of continuous operation, Miller and Urey found that many of the molecules common to living things had been produced. These included sugars, lipids, and amino acids, which are the building blocks of proteins. The experiment has been modified using different gases in different compositions and employing alternative energy sources. All variations lead to essentially the same result, indicating that life's molecules are quite robust: given a broad set of initial conditions, these molecules will arise spontaneously.

5. There was nothing particularly energy efficient about steam engines of the 1800s. Recall that Carnot proved maximum engine efficiency depends on the temperatures of the two heat reservoirs between which the engine operates. The greater the difference, the more efficient the engine can be. In the case of a steam engine, the boiler is the high-temperature reservoir and the condenser, low temperature. With nineteenth-century materials, the highest temperatures possible in a good-size steam boiler were on the order of a few hundred degrees Celsius, and the condenser could get no colder than ambient temperatures (the temperature of the environment surrounding the condenser). So the best possible temperature differences were just a few hundred degrees. When you consider that the combustion gases from the coal fueling the steam engine are 1,000+ degrees hotter than ambient temperature, it's clear there is a lot of wasted energy. If one could bypass the steam altogether and use the combustion products from coal to push pistons, engine efficiencies could really be upped. The difficulty is that lumps of coal burn too slowly to generate any real power this way. Although coal dust burns fast and can be used to power engines directly, it suffers from copious other problems that, in the late eighteenth and early nineteenth centuries, made it an impractical fuel. Rudolf Diesel designed an engine fueled by coal dust but soon converted it to run on the distillate from crude oil that bears his name, Diesel. Today, researchers are experimenting with coal dust as a fuel to power turbine engines.

6. Evidence exists that 1.5 billion years ago, at Oklo in Gabon in Africa, there was a natural nuclear reactor that formed when a uranium-rich mineral deposit became inundated with neutron-moderating groundwater. A moderator slows neutrons down and increases the likelihood that it will induce fission. The water moderator would boil away as the reaction increased, slowing it down and preventing a meltdown. The fission reaction was sustained for hundreds of thousands of years.

7. A possible consequence of Rachel Carson's book was the spread of malaria. In a report titled "A Review of the Twenty Greatest Unfounded Health Scares of Present Times" from the American Council on Science and Health, the ban on DDT holds a prominent place. ACSH states that the benefits of DDT's power to kill insects that may carry diseases that threaten humans outweigh the chemical's effects on wildlife and humans. In addition, ACSH questions whether DDT affects wildlife and humans adversely.

   It is believed that malaria afflicts between 300 and 500 million people every year, causing up to 2.7 million deaths, mainly among children under five years. In the 1960s, DDT was the foremost weapon battling this scourge. Through its use, Latin America was rid of the mosquitoes carrying the malaria parasite. Since the DDT ban, malaria has resurfaced. Though the evidence is far from establishing a cause-and-effect relationship, many blame the DDT ban for the resurgence of this dread disease.

References

CHAPTER 1

Roberts, Jeffrey W. *The Drinking Man's Diet*. San Francisco: Cameron and Company, 2003. First published 1964.

CHAPTERS 2–5

Von Baeyer, Hans Christian. *Warmth Disperses and Time Passes: The History of Heat*. New York: Modern Library, 1999.
What does energy really mean? *http://physicsweb.org/articles/world/15/7/2/1*.
Perpetual motion. *http://www.trivia-library.com/b/history-and-quest-for-perpetual-motion-1191-to-1500.htm*.
Plenio, M. B., and V. Vitelli. "The Physics of Forgetting: Landauer's Erasure

Principle and Information Theory." *Contemporary Physics* 42 (2001): 25–60.

CHAPTERS 6–9

DeGrasse Tyson, Neil, and Donald Goldsmith. *Origins: Fourteen Billion Years of Cosmic Evolution.* New York: W. W. Norton, 2004.

Johnson, Calvin W. "Goldilocks and the Three Planets." *http://www.phys.lsu.edu/faculty/cjohnson/climate.html.*

Miller, Andrew. *Oxygen.* New York: Harvest Books, 2003.

Woo, Janet. "The Beginning of Life and Amphiphillic Molecules." *http://www.scq.ubc.ca/?p=505.*

Segre, Daniel Dafna Ben Eli, David W. Deamer, and Doron Lancet. "The Lipid World." *Origins of Life and Evolution of the Biosphere* 31 (2001): 119–45.

Introduction to cyanobacteria. *http://www.ucmp.berkeley.edu/bacteria/cyanointro.html.*

Deffeyes, Kenneth S. *Beyond Oil: The View from Hubbert's Peak.* New York: Hill and Wang, 2005.

Petroleum formation. *http://tesla.jcu.edu.au/schools/earth/EA1002/Energy/EnergyPetroleum.html.*

CHAPTER 10

Stanley, Steven M. *Children of the Ice Age: How a Global Catastrophe Allowed Humans to Evolve.* New York: Harmony Books, 1996.

Diamond, Jared M. *The Third Chimpanzee: The Evolution and Future of the Human Animal.* New York: Harper Perennial, 2006.

Falk, Dean. *Braindance: New Discoveries about Human Origins and Brain Evolution.* Gainesville: University Press of Florida, 2004.

Aiello, L. C. and P. Wheeler. "The Expensive-Tissue Hypothesis: The Brain and the Digestive System in Human and Primate Evolution." *Current Anthropology* 36 (1995): 199–221.

Leonard, William R., Marcia L. Robertson, J. Josh Snodgrass, Christopher W. Kuzawa. "Metabolic Correlates of Hominid Brain Evolution." *Comparative Biochemistry and Physiology Part A* 136 (2003): 5–15.

Bramble, Dennis M., and Daniel E. Lieberman. "Endurance Running and the Evolution of Homo." *Nature* 432 (2004): 345–52.

Carrier, David R. "The Energetic Paradox of Human Running and Hominid Evolution." *Current Anthropology* 25 (1984): 483–95.

MacLarnon, Ann, and Gwen Hewitt. "Increased Breathing Control: Another Factor in the Evolution of Human Language." *Evolutionary Anthropology* 13 (2004): 181–97.

Patterson, J. H. *The Man-Eaters of Tsaro.* New York: St. Martin's, 1986.

Shreeve, James. *The Neandertal Enigma: Solving the Mystery of Modern Human Origins.* New York: Harper Perennial, 1996.

CHAPTER 11

Smil, Vaclay. *Energy in World History.* Boulder, Colo.: Westview Press, 1994.

Diamond, Jared M. *Guns, Germs, and Steel: The Fates of Human Societies.* New York: W. W. Norton, 1999.

The development of agriculture. *http://www.unl.edu/rhames/courses/orig_agri_tur.html.*

CHAPTERS 12–13

Diamond, Jared. *Collapse: How Societies Choose to Fail or Succeed.* New York: Penguin, 2005.

History of London. *http://www.britainexpress.com/London/medieval-london.htm.*

Cowen, Richard. "Exploiting the Earth," unpublished ms., chap. 10–11. *http://www-geology.ucdavis.edu/~cowen/~GEL115/index.html.*

CHAPTER 14

Yergen, Daniel. *The Prize: The Epic Quest for Oil, Money and Power.* New York: Free Press, 1993.

CHAPTER 15

Niagara Falls, History of Power. *http://www.niagarafrontier.com/power.html.*

Radioluminescent Paint. *http://www.orau.org/ptp/collection/radioluminescent/radioluminescentinfo.htm*

CHAPTER 16

Love Canal. *http://www.epa.gov/history/topics/lovecanal/index.htm.*
Oil industry statistics. *http://www.gravmag.com/oil.html.*
Solar dimming. *http://news.bbc.co.uk/2/hi/science/nature/4171591.stm.*
Brook, Edward J. "Tiny Bubbles Tell All." *Science* 310 (25 November 2005): 1285–87.
Siegenthaler, Urs, et al. "Stable Carbon Cycle—Climate Relationship During the Late Pleistocene." *Science* 310 (25 November 2005): 1313–17.
Antarctic ice core summary. *http://www.realclimate.org/index.php?p=221.*

CHAPTERS 17–18

Fuel cells. *http://www.benwiens.com/energy4.html#energy1.7.*
Tesla's electric car. *http://www.teslamotors.com/.*
Another vision of our energy future. *http://ergosphere.blogspot.com/2004_11_01_ergosphere_archive.html.*

Index

Page numbers in **boldface** refer to illustrations.

AC current, 198, 199
Acheson, Dean, 188
Africa
  Australopithecus, 124–126, 127–128
  Homo erectus and *Homo habilis,*
    133–136
  Homo sapiens, 139
  predators and human evolution,
    126–128
  San (Bushmen) of the Kalahari,
    132–133
agriculture, 145–150
Aluminum Company of America
  (ALCOA), 201–202

ammonia, 107
ammonia engine, 28–29
amphipathic molecules, 108–110, 111
Anasazi, xiv–xv
Anglo-Persian Oil Company, 187
animal power, 147–150
animals, evolution of, 114
"Another Ice Age?" *(Time),* 221–222
anoxic conditions, 115
Antarctic ice cores, 222–223
Arab oil embargoes, 190–191
Arctic National Wildlife Refuge
  (ANWR), xvii
Arctic Ocean, 123–124

Argand, Aimee, 173
Armstrong, Lance, 50
art, 139–140
Astor, John, 201
atmospheres, 100–103, 113
atoms
  basic particles, 81–83
  Brownian motion, 55–56
  distribution of, 57–59
  energy changes, 83
  evolution of, 85–87
  existence of, 22, 25, 53–54, 55–56
  nucleus, 81–82
  orbitals, 82, 83
Australopithecus, 124–133
automobiles
  electric, 252–256
  gasoline, 182–183
  hybrids, 253–254
  hydrogen fuel cells, 16, 26–27
  power and, 10

bacteria, 112–114
Baku, 184–185
basal metabolic rate (BMR), 4
batteries, 253, 254–255
Beaumont, Huntington, 169
Beck, Eckardt C., 233
Berzelius, Jöns Jakob, 18
Bhopal disaster (India), 234
Big Bang, 64, 71, 80, 89, 91
bin Laden, Osama, xviii
biomass, 260–261
bipedal stance/locomotion, 127–128,
  129
Bissell, George, 175, 176, 177
bitumen, 175–176
Black Sea, 115, 116
Black Sunday, 247–248
Boltzmann, Ludwig, 52–53, 55–56,
  56–59, 264
Boltzmann Constant, 58–59, 264

boundary, system, 39–40
Brain, Charles, 125
*Braindance: New Discoveries about
  Human Origins and Brain
  Evolution,* 128
brain size, 131, 135–136, 136–138
brainwork, 4–8, 63–64
brewing industry, 166
Brindley, James, 169–170
British Petroleum (BP), 187
Brownian motion, 55–56
Brush, Charles, 200
Bryson, Reid A., 221, 224
Buffalo (New York), 201
bullet trains, 253
Burke, Arnold, 15
Burton, William Merriam, 183–184
Bush, George W., xvi–xviii, 16, 26–27,
  192–193, 217
Byrn, Edward, 237

Calder Hall power station, 208
caloric theory of heat, 18–19, 21, 32, 33
Cameron, Robert, 3–4
canals, 169–170
cancer, 166
candles, 173
candy bars, 25
cannonballs (gunstones), 158–159
cannons, 20–21, 158–159
caprock, 117
carbon, 89, 90
carbon capture and storage (CCS), 259,
  260
carbon cycle, 103–105, 221
carbon dioxide (CO_2), 101–105,
  220–221, 259–260
carbonic acid, 103
Carey, Hugh, 233
carnivores, 125
Carnot, Sadi, 31–35, 36–37
Carson, Rachel, 231, 267n7

Carter, Jimmy, 191, 233
Caspian Sea, 184–185
catalase, 114
CCS (carbon capture and storage), 259, 260
cells, 110–112
cellulose, 114
Central American Seaway, 123
cereals, 147
chain reaction, 206
Challenger space shuttle explosion, 234
Chanute, Octave, 236–237
chaos theory, 238–239
charcoal, 156–159, 161–162
charge density, 83
Charlotte Dundas (steam-tug), 170
chemical energy, 9–10. *See also specific types*
chemical industry, 232–234
chemical reactions, 93–98
Chernobyl nuclear accident, 210–211
Children of the Ice Age (Stanley), 122
China, 188, 224–225, 259
The China Syndrome, 232
choices, 49
Churchill, Winston, 185–187, 220
Clark, Maurice, 179
clathrates, 120
Clausius, Rudolf, 37–38, 43–44, 47
climate change
  global warming, xiv, xvi, 102, 215–216, 220–223, 241–242
  ice ages, 124, 125–126, 221–222
  solar dimming, 223–225
  weather extremes, 225
coal
  ancient use of, 163
  coke production, 166–167
  electricity production, 203, 259–260
  formation of, 118–119
  iron and steel industry, 167
  kerosene from, 173–174

mining, 167–168
types of, 119
use in medieval England, 163–166
use in U.S., 170–171
coal dust, 266n5
COBE (Cosmic Background Explorer) satellite, 265n3
Cohen, Bernard, 256n
coke, 166–167
cold fusion, 245–246
Cold War, 249–250
Colorado oil shale project, 247–248
Columbia River dams, 203
communications networks, 60
complexity, 238–239
conservation of energy. *See* First Law of Thermodynamics
Cook, William, 197
cooling mechanisms. *See* heat tolerance
coral polyps, 228
cosmic microwave background radiation, 265n2, 265n3
Council on Foreign Relations, 218–219
Cowen, Richard, 160, 165
cranial blood flow, 127–131
creativity, 139–140. *See also* imagination
cultural development and energy, xiv–xv, 8
current, 198–199
cyanobacteria, 113–114

dams, 202–203
Darby, Abraham, 167
Darwin, Charles, 49–50, 62–63
Da Vinci, Leonardo, 15
DC current, 197, 199
DDT, 231, 267n7
deforestation, xv–xvi, 153–155, 160–162, 165–166
density variations, 87
dependent variables, 41–43
deuterium, 81, 86

Diesel, Rudolf, 266n5
diets, 3–5, 10–11
Dietz, Michael, 173–174
diploic veins, 129
discretionary hours, 144–145
draft animals, 147–150
Drake, Edwin L., 174, 176–177
Drake, Frank, 227
The Drinking Man's Diet (Cameron), 3–4
Dundas, Lord, 170
dust, 92–93
dynamos, 23

Earth, 101–104
Easter Islanders, xiv–xv
ecology, 231
Eddington, Arthur, 240–241
Edison, Thomas Alva, 181, 197
efficiency
    as measure of fitness, 50–51, 62–64
    of processes, 43–44
    steam engines, 31–34
Egypt, 189–190
Einstein, Albert, 25, 55–56, 122, 240–241
electricity, 194–211
    from coal, 203, 259–260
    current generation, 195–196
    dams, 202–203
    lights, 181–182, 197
    motors, 195
    Niagara Falls, 199–202
    solar energy, 256–257
    telegraphy, 196–197
    transmission, 197–199
    for transportation, 252–256
    wind energy, 256
electric transportation, 252–256
electrochemical industries, 201–202
electrons, 80–83, 85–87
electrostatic force, 81, 87–88, 93

An Elementary Treatise on Chemistry (Lavoisier), 18
Element Range (universe's energy ride), 73, 76, 77
elephants, 128
emissary foramina, 130
emissary veins, 129
energy conservation. *See* First Law of Thermodynamics
energy dependence, xii–xiii, xiii–xiv, xvii, 8, 216–220
energy diet, 10–11, 13, 51, 242. *See also* energy future
energy dilution, 44–51, 58–59
energy forms, 16–18, 29–30
energy future
    ambivalence about, 9–10
    choosing, xvii–xviii
    electricity generation, 256–260
    electric transportation, 252–256
    imagining the future, xviii, 11, 251–252, 262
    United States as energy supplier to the world, 261–262
energy policy, xiii, xvi–xviii, 246, 248–249
energy ride of the universe, 71–79
England
    charcoal production and use, 155–162
    coal mining, 168
    coal transportation, 168–170
    coke production, 166–167
    Medieval coal use, 163–166
    nuclear reactors, 208
    World War I oil supply, 185–187
entropy
    Boltzmann Constant and distribution of molecules, 55–59, 264
    choice and, 49–51
    Clausius and, 37–38, 43–46
    as information, 60–64
    irreversible processes, 44–49

order and, 59
time and, 49
EPICA (European Project for Ice Coring in Antarctica), 223
ethanol, 260–261
Eureka moments, 6–7
Europe, energy crises, xv–xvi
European Project for Ice Coring in Antarctica (EPICA), 223
Eurostar, 253
Evans, Robley, 204–205
Evelyn, John, 165
Evershed, Thomas, 200
expansion, 19
explosions, 67–69, 93–94
explosive mixtures, 95–98
Exxon, 247–248
Exxon Valdez oil spill, 234

Falk, Dean, 128, 130
Faraday, Michael, 195
Federal Clean Air Act (1974), 224
Fertile Crescent, 145–146
fire, 136
First Law of Thermodynamics (conservation of energy), 13–27
    caloric theory and, 18–19, 21
    forms of energy, 13–14, 16–17, 61
    heat and motion, 21–22
    mass conservation, 26
    mechanical equivalent of heat, 22–25
    perpetual motion machines, 14–16, 26–27
    and the universe's energy ride, 73
fission, 204–207, 267n6
fitness, 62–64
Flagler, Henry, 180
Fleischer, Ari, 217
Fleischmann, Martin, 245–246
Ford, Henry, 182
forests. *See* wood
forgetting, 52, 61

fossil fuels, formation of, 115–120
Foster, Craig, 132
France, 138, 208
Franch, Guido, 15
friction, 19
fuel cell vehicles, 254–255
"Fumifigium" (Evelyn), 165
fusion reactions, 88–91, 204, 245–246

galaxies, 76–77
Gamgee, John, 28–29, 37
gas can bomb, 67–69, 93–94
gasoline, 180, 182–184
Gesner, Abraham, 173
Glen Canyon Dam, 203
Global Ocean Conveyor, 124, 216
global warming, xiv, xvi, 102, 215–216, 220–223, 241–242
glycerin and potassium permanganate, 96–98
grain cultivation, 146–147
gravity, 48, 74, 87, 90, 91, 99
Great Barrier Reef, 228
The Great Dance: A Hunter's Story (Foster), 132
Great Energy Rule, 144–150
Greek fire, 176
greenhouse gases, 102, 220–223
Green River Basin, 118, 247–248
gunpowder, 95
gunstones, 158–159

Harrison, William, 165–166
Hart, Alexander, 170
heat, 17–27
    caloric theory of, 18–19, 21, 32, 33
    Carnot and steam engine efficiency, 31–34, 36–37
    as empty calories, 30
    expansion and, 19
    frictional, 19
    mechanical equivalent of, 22–25

heat *(continued)*
  motion and, 21–22, 25–26
  temperature *vs.,* 25–26
heat death, 44–51
heat tolerance
  cranial blood flow, 127–131
  respiratory control, 135
  sweat glands, 134–135
heavy oil, 117
helium, 86–91, 99
Helium Valley (universe's energy ride),
  76
Helmholtz, Hermann von, 25
Hill, Gladwin, 231
Homo erectus, 131, 133–138
Homo habilis, 130–133
Homo sapiens, 138–140
Honnecourt, Villard de, 14–15
Hooker Chemical Company, 232–233,
  239–240
Hoover Dam, 203
horses, 148–150
human evolution
  agriculture, 145–150
  Australopithecus, 124–133
  bipedal stance/locomotion, 127–128,
    129
  brain size, 131, 135–136, 136–138
  cooking food, 138
  cranial blood flow, 127–131
  creativity, 139–140
  discretionary hours, 144–145
  heat tolerance, 127–131, 134–136
  Homo erectus, 131, 133–138
  Homo habilis, 130–133
  Homo sapiens, 138–140
  hunting and tracking skills, 131–134
  predators and, 125, 126–128
  respiratory control, 135
  social groups, 145
  sweat glands, 134–135
  tools, 136, 139

hunting and tracking skills, 131–134
Hussein, Saddam, xvi–xviii, 191–192,
  192–193
hybrid cars, 253
hydrates, 120
hydroelectric energy, 9, 199–203
hydrogen, 87–91, 99
hydrogen fuel cells, 16, 26–27
Hydrogen Ravine (universe's energy
  ride), 74, 75–76, 78–79, 80
hydrogen sulfide, 107
hydrophilic molecules, 107, 109, 110
hydrophobic molecules, 107–109, 110
hyenas, 126–127

ice ages, 124, 125–126, 221–222
ice cores, 222–223
Idaho nuclear reactor, 208–209
IGCC (integrated gasification combined
  cycle), 259–260
Iliad (Homer), 176
imagination
  as brainwork, 6–8
  and energy future, xviii, 251, 262
  fuel for, xi–xii, xiv–xv, 9, 140
  knowledge acquisition, 63, 143
  unique to humans, 229
  wishing *vs.,* 245–246
independent variables, 41–43
India, Bhopal disaster, 234
Industrial Revolution, 167, 168, 170
information theory, 60–64
infrared light, 101
integrated circuits, 257–258
integrated gasification combined cycle
  (IGCC), 259–260
Iran, 219–220
Iraq, xvi–xviii, 191–193
iron, 89, 90, 91
iron and steel industries, 157–159, 167
Iron Basin (universe's energy ride), 77
irreversible processes, 44–49

Isherwood, B. F., 29, 37
Islamic radical movements, 217,
    218–220
isotopes, 81, 206
Israel, 189–191
Isthmus of Panama, 122–124

Japan, 8, 187–189
Jekyll and Hyde mentality, 9–10
Johnson, Donald, 125
Joncairs, Daniel, 199
joule, 24–25
Joule, James Prescott, 22–25
Jurassic Park, 238

Kalahari Desert, 126, 132–133
Kelvin, William Thomson, Lord,
    22–23
Kennedy, John F., xvii–xviii, 262
kerosene, 173–174, 178–181, 183
Khamenei, Ali, Ayatollah, 219–220
kinetic energy (motion), 13–14, 16–17
knowledge, 60–64, 143–144
Kuwait, 191–192

Lavoisier, Antoine, 18
life
    amphipathic molecules, 108–110, **111**
    building blocks of, 78
    cells, 110–112
    efficiency and survival, 62–64
    Miller and Urey's experiment on
        origins of, 265n4
    photosynthesis and respiration,
        112–115
    water and, 106–108
light
    campfires, 172
    candles, 173
    electric, 181–182, 197
    and greenhouse gasses, 101–102
    kerosene, 173–174, 178–181

olive and whale oil lamps, 172–173
    spectrum, 83–85
limestone, 104, 153
lions, 126–127
lipids, 110–112
logarithms, 263n1
London. *See* England
Love Canal, 232–233, 239–240
lubricating oils, 183
Lucy, 125

Mach, Ernst, 55
magnetic induction, 196
malaria, 267n7
The Man Eaters of Tsavo (Patterson), 127
Mars, 101, 102, 103, 105
Maxwell, James, 53–55, 56
Mayan civilization, xiv–xv, 151–155
melt down, 207–208
meningeal veins, 129
metabolism, 12–13
methane, 119–120, 183
methane hydrates, 120
micelles, 110–111
Middle East, xiii, 117, 185, 189–193,
    217–220
military-industrial complex, 250–251
Miller, Stanley, 265n4
Million Solar Roofs Initiative, 257
Molecular Downs (universe's energy
    ride), 77–78
molecular reactions, 93–98
molecules
    distribution, 56–59
    existence of, 22, 25, 54
    formation of, 93
    planets and, 98–101
    speed of, 54–55
    and water, 107–110
Morgan, J. P., 201
Morse, Samuel, 197
motion (kinetic energy), 13–14, 16–17

Mount Genesis (universe's energy ride), 72, 74–75
muscles, energy consumption of, 4–5

NABISCO (National Biscuit Company), 202
naphthas, 183
natural gas, 119–120
natural selection, 49–50, 63
Navy, U.S., 29, 250–251
The Neandertal Enigma: Solving the Mystery of Modern Human Origins (Shreeve), 140
neon lights, 84–85
neutrons, 80–81, 85–87
Newcomen, Thomas, 18, 168
Newton, Isaac, 17, 30, 48
New York Times, 231, 233
Niagara Electro Chemical Company, 202
Niagara Falls, 199–202
nickel, 89, 90, 91
Nixon, Richard, 190, 191, 224
Nobel, Robert and Ludvig, 184
North and South America, 122–123
nuclear accidents, 208–211, 232
nuclear energy, 9, 204–207, 208–211, 256, 258
Nuclear Power 2010, 258
nuclear-powered ships, 250–251
nuclear reactions, 93–94
nuclear reactors, 206–211, 250–251, 261, 267n6
nuclear weapons, 93–94, 207
nucleus, 81–82

Occidental Chemical, 239–240
oceanic circulation, 123–124
oil
  current production, 216
  current usage, xii
  distillation fractions, 175, 180, 182
  drilling for, in Pennsylvania, 174–178

Exxon Valdez oil spill, 234
  formation, 116–117
  gasoline, 182–184
  internal combustion engines, 182–184
  kerosene for illumination, 178–181
  locations of, 184–185
  Middle East, xiii, 117, 175–176, 185, 189–193, 217–220
  as Native American medicine, 174
  as weapon, 176
oil embargoes
  Arab oil embargoes, xii–xiii, 191–192
  on Iraqi oil, 192
  U.S. embargo against Japan (World War II), 8, 185–189
oil lamps, 172–173
oil shale, 117–119, 247–248
oil window, 116, 119
"On a Different Form of the Second Law of Thermodynamics" (Clausius), 37–38
Oort cloud, 106
OPEC, 189
orbitals, 82, 83, 93
order, 59
Organization of Petroleum Exporting Countries (OPEC), 189
Out of Africa model, 139
oxygen, 90–91, 112–114

Panama isthmus, 122–124
parsimony, principle of, 122
particulates, 223–225
Patterson, J. H., 127
Pearl Harbor, 187–189
peat bogs, 118–119
Pennsylvania coal and oil, 171, 174–178
Perky, Henry D., 202
perpetual motion machines, 14–16, 26–27
pesticides, 231
phase space, 72

Philip, king of Spain, 161
Photon Flats (universe's energy ride), 75, 76, 80
photons, 83
photosynthesis, 112–115
Pioneer 10 and *11* spacecraft, 226–227, 228–229, 230
planetesimals, 100
planets, formation of, 92, 98–101
plants, evolution of, 114
Platt, Henry, 166
plutonium, 206, 207
pollution, 163–166, 223–225. *See also* climate change
Pons, Stanley, 245–246
population, 149–150
potassium permanganate and glycerin, 96–98
potential energy, 16–17
power, 10
predation, 125, 126–128, 131–134
The Prize (Yergen), 178–179
processes, 42–43
protons, 80–82, 85–87
protostars, 87

quantum mechanics, 82
quicklime, 153–154

radiator veins, 129–130
radium, 204–205
railroads, 169, 170, 253
rattlesnakes, 101
Reagan, Ronald, xiii, 191
Red Sea, 115
red shift, 265n2
red supergiants, 90–91
Reflections on the Motive Power of Fire (Carnot), 32
repeatability, 40–41
repulsive electrostatic barrier, 87, 88
resistance, 198–199

respiration, 114–115
respiratory control, 135
reversible processes, 34–37, 42–44, **43,** 47
Rockefeller, John D., 179–181
Rock Island Dam, 203
Rodger (lab assistant), 96–98
roller coasters, 69–71
Roosevelt, Franklin D., 188
Royal Dutch Oil Company, 185
Rumford, Benjamin Thompson, Count, 19–21
rust, 113

Sagan, Carl, 227
Sahara Desert, 126
San (Bushmen), 132–133
Saudi Arabia, 190–191, 217–219
Schoellkopf, Jacob, 200
science and technology
    complexity and, 238–239
    confidence in, xvii–xviii, 230
    human technological immaturity, 235–236, 239
    inability to predict path of, 236–237
    inevitability of mistakes, 239–240
    measurements and repeatability, 39–43
    pessimism about, 231–235
    uncertainty in, 240–241
Second Law of Thermodynamics, 29–38
    Carnot and steam engine efficiency, 31–34, 36–37
    consequences of, 47–51, 62–64
    entropy, 37–38, 43–46
    quality of energy, 29–30
    reversibility, 34–37, 44–49
    roller coasters and, 70–71
    statements of, 30, 33, 34, 37, 44
    and the universe's energy ride, 74
Seneca Oil, 174
Shannon, Claude, 60
Shea, Nina, 219

Sherman Antitrust Act, 181
shipbuilding industry, 159–161
Shippingport reactor, 208
Shreeve, James, 140
Silent Spring (Carson), 231
Silliman, Benjamin, Jr., 175
SL-1 reactor (Idaho), 208–209
Smith, "Uncle Billy," 177
soap, 109
social groups, 145
solar dimming, 223–225
solar energy, 256–257, 261
solar spectrum, 85
solar system, 98–101
Somerset, Duke of, 160
spectrometers, 84
spectrum, 84–85
Standard Oil, 181, 183–184
Standard Works, 180
Stanhill, Gerry, 223–224
Stanley, Steven, 122
starch, 114
stars, 76–77, 85–88, 90–91, 99–100
state variables, 41–43
steam engines, 18, 31–33, 168, 170,
    266n5
steamships, 170
steel and iron industries, 157–159, 167
steps (universe's energy ride), 73
Sterkfontein Cave (South Africa), 125
stoves, development of, 19, 235–236
streetlights, 83–84
strong force, 81, 86, 87, 88
stucco, 153–154
subduction, 104
sugar, 114
sulfur, 166, 167
sulfur dioxide, 166
sulfuric acid, 166
Sumatra, 185
Sun, formation of, 85, 99–100
supergiants, 90–91

supernovas, 91
surroundings, 39–40
survival of the fittest, 49–50, 62–64
sweat glands, 134–135
syngas, 260
Syria, 189–190
system, 39–40
system boundary, 40

tar sands, 117
technology. *See* science and technology
tectonic plates, 104, 123
telegraphy, 196–197
termite mounds, 228
Terra Amata (France), 138
Tesla, Nikola, 197–198
Tesla Roadster, 253
Texas, 185
thermodynamics. *See also* First Law of
    Thermodynamics; Second Law of
    Thermodynamics
  efficiency as measure of fitness, 50–51,
    63
  terminology of, 39–43
  use of term, 21
thinking man's diet, 4–5
thinking man's energy diet, 10–11, 13,
    51, 242. *See also* energy future
third rails, 255–256
Three Mile Island nuclear accident,
    209–210, 232
Tikal (Yucatán Peninsula), 151–155
time, 48–49
Time, 221
Titanium Alloy Manufacturing
    Company, 202
tools, 136, 139
total energy, 17
Townsend, James, 175, 176, 177
Trevithick, Richard, 170
triple-alpha process, 88–89
tritium, 81, 86

turpentine, 108–109
TWA Flight 800, 95

U-235, 207
Union Carbide, 234
United States
  current oil consumption, xii, 216
  defense policy, 249–250
  energy and cultural development in, 8
  energy dependence, xii–xiii, xiii–xiv,
    xvii, 8, 216–220
  energy policy, xiii, xvi–xviii, 246,
    248–249
  as energy supplier to the world,
    261–262
  nuclear power, 208–210
  oil as weapon in World War II,
    187–189
  whaling industry, 173
universe
  atoms, formation of, 80–87
  Big Bang, 64, 71, 80, 89, 91
  elements, formation of, 87–91
  energy ride of, 71–79
  heat death of, 44–51
  planets, formation of, 98–101
uranium, 206, 207
Urey, Harold C., 265n4

Vail, Alfred, 197
Vanderbilt, William, 201
variables, 41–43
Venus, 101, 102–103, 104–105
vesicles, 110–112
visible light, 83–84, 101
Vogelherd (Germany), 139
volcanoes, 104, 107
voltage, 198–199
Von Baeyer, Hans Christian, 31
von Neumann, John, 60
Voyager 1 and 2 spacecraft, 226,
  227–229, 230

wagon-ways, 169
Wahhabi madrassas, 217, 218–219
Washington Post, 219
watch dial painters, 204–205
water, 101, 106–112
water vapor, 102
Watt, James, 31
weather extremes, 225
Westinghouse, George, 201
whale oil, 173
"What if?" questions, 6–7, 229–230.
  See also imagination
wheat, 146
Wheatstone, Charles, 197
wind energy, 256
Windscale nuclear site, 208
wishing, 245–246
wood, 8–9
  charcoal production, 156–157,
    161–162
  deforestation, xv–xvi, 153–155,
    160–162, 165–166
  industrial use of charcoal, 157–159
  production of quicklime, 153–155
  wood-burning industries, xii,
    159–161, 166
work, xv, 4–8
World Health Organization (WHO), 224
World War I, 185–187
World War II, 187–189

Yamamoto, Isoroku, 189
Yergen, Daniel, 178–179
Yom Kippur invasion, 189–190

zeromotor, 29
Zijlker, Aeilko Jans, 185
Zimara, Mark Anthony, 15

About the Author

DR. MARK E. EBERHART is a fifth-generation native of Colorado. He grew up in Denver where he attended public school. One institution where he sought an education was Stedman Elementary School, which was named after his great-great-grandfather who was among the first physicians in Denver—treating patients at the peak of the Colorado gold rush. During the summers Mark would travel to Salem, Oregon, to be with his father and stepfamily. At age sixteen Mark enrolled at the University of Colorado, where he received both a B.S. with majors in chemistry and applied mathematics and an M.S. in physical biochemistry. In 1979 he applied and was accepted to MIT as a Ph.D. candidate studying materials science and engineering.

At the same time Mark was packing his belongings in preparation for the move to Boston, the Iranian revolution was in full swing and gasoline in the United States was in short supply. News accounts of a gas-starved public waiting in service-station lines for hours were common. Undeterred by the uncertainty, Mark placed six five-gallon cans full of gas next to his other possessions in the rented U-Haul and headed east. This trip seeded an interest in energy science and policy that never faded.

Four years later Mark received his Ph.D. and was one of a handful of scientists attempting to understand fracture at the quantum mechanical level. The pursuit took him from MIT to Los Alamos National Laboratory and from there to the premier university for engineering in the Rocky Mountains, the Colorado School of Mines, where he is now a professor teaching chemistry and materials science.

Mark lives in Denver, just a few miles from where he grew up, and daily rides his bike up the 2,000-foot climb from his office on the CSM campus to the top of Lookout Mountain; he keeps careful records of the energy he expends making the climb, noting each year that as his conditioning improves, the energy wasted on this ride declines. He is a consultant to NOVA and a popular speaker—giving presentations as diverse as the role of science in society, to Boston's Great Molasses Disaster and other failures that have shaped engineering, science, and technology.